– IN THE MOMENT –

The Butterfly

Flights of Enchantment

Nigel Andrew

Saraband

Published by Saraband
3 Clairmont Gardens
Glasgow, G3 7LW

www.saraband.net

ISBN: 9781916812338

Printed and bound in Great Britain by Clays Ltd, Elcograf S.p.A.

1 2 3 4 5 6 7 8 9 10

In memory of my father,
E.R. 'Bob' Andrew

The world of the living contains enough marvels and mysteries as it is; marvels and mysteries acting upon our emotions and intelligence in ways so inexplicable that it would almost justify the conception of life as an enchanted state.
Joseph Conrad, preface to *The Shadow-Line*

The world is so full of a number of things,
I'm sure we should all be as happy as kings.
Robert Louis Stevenson,
A Child's Garden of Verse

Contents

Contents

Part I

A Personal Flight

My Butterfly Life

Dappled sunlight falls on a classic English woodland ride; the wide grass bridleway is fringed with cow parsley, willow herb and campion, with goat willow and hazel further back, and wild honeysuckle draping the trees that rise above the lower shrubs. The brambles are in full flower, and they are alive with butterflies: mostly Gatekeepers and Meadow Browns, but among them, and gliding elegantly along the ride, in and out of the beams of sunlight – now you see them, now you don't – are White Admirals, ethereal creatures of near-black and white, much larger, but seeming almost substanceless in flight. And then one or two of them settle to take nectar from the bramble flowers, and fold their wings to show the exquisite markings of their undersides: bands of white across a chestnut ground spotted with rows of near-black dots, the outer edges delicately lined with dots of alternating black and white, and near the body a wash of palest blue-grey. I think I have never seen anything so beautiful ...

I am there, on this enchanted ride, with my brother and my father. We all have butterfly nets in our hands. We will very probably take a swipe and try to catch one of these beauties as it feeds or basks, or give hopeless chase after one in full flight along the ride. We shan't be killing or setting them; those days are coming to an end, and our nets are already beginning to look like relics of another age.

Where was this ride? Perhaps in the New Forest, perhaps in Ruislip woods, that magical woodland which, improbably, you could walk into from the edge of Ruislip Lido, where sunbathers enjoyed the nearest thing to the seaside that Middlesex had to offer. It hardly matters where it was: this scene – memory or dream, or something of both – has remained with me through life as an epitome of the ecstatic joy that the pursuit of butterflies, with or without a net, can give, and as one of my fondest memories of being with my father. I have ever since associated the White Admiral – a species I make sure to see every summer – with that scene, that golden moment of joy.

It all began long before I was born – in my father's post-Edwardian boyhood. Like every boy of his time, growing up in the late teens and twenties of the last century, my father and his three brothers led a life of what now seems idyllic freedom, ranging at large over the countryside, on foot, on bicycles and, later, on motorbikes. This free-ranging life led naturally, as it did for so many boys (and quite a few girls) in those days, to an abiding interest in wildlife in all its forms – especially birds and butterflies. Bird-nesting – seeking out birds' nests and taking eggs, which were then 'blown' and preserved in (usually small) collections – was standard practice, as it was even into my own boyhood in the late 1950s and early 1960s. (In defence of this, I can attest that it encouraged close study of bird behaviour, demanded a very sharp eye and good tree-climbing skills, and offered a unique insight into the birds' world; also it took no toll on most birds, provided you were discreet about it, took only one egg from a clutch, and were careful not to alarm the parent birds.)

Likewise, netting butterflies, and killing and setting some of them, was a very widespread pursuit among the boys and young men of my father's generation. Most of his early years were spent in or near the Peak District of Derbyshire, which is not prime butterfly country, but such was his love of butterflies that, as a young man, he would make frequent holiday excursions to the New Forest, where many of our most spectacular species – Silver-Washed and other Fritillaries, White Admirals, even Purple Emperors – flew in abundance. The more he chased butterflies, the more my father became interested in their life cycles and other aspects of their biology – to the point where he fancied he would like to study entomology at university. When he informed his father of this ambition, he was promptly told not to talk such nonsense, and that he'd start in 'the works' next week. And so he embarked (like so many in his family) on a career in engineering. But his love of butterflies never left him, and when he became a father he happily passed it on to his two sons. It was one thing he and I always had in common, even when, in the course of my turbulent and overlong adolescence, relations between us became tense and difficult. We could always talk butterflies.

And now the scene shifts to another little paradise, this one on the coast of Kent, near Deal. Here, in a whitewashed cottage with a pungent, unidentifiable smell (gas? damp? both?), some of the happiest holidays of my boyhood were spent. It was here that my father introduced me and my brother to the various tribes and species of butterflies, their different tastes and habits, and, of course, the art of netting them with our baggy tartalan nets. Having caught one – and many, most got away – we would attempt, with varying degrees of success, the tricky task of transferring it from the net into a round, glass-topped metal box, in which, once a hand had been placed over the glass for a minute, it would obligingly fall asleep, allowing us fully to admire the beauty of its markings. Then we would lift the lid, and off it would fly, none the worse for its brief captivity. Sometimes, before taking off, it would walk groggily onto an outstretched finger to recover its energy. On a particular patch of waste ground in front of the cottage, which we haunted whenever the sun was out, we would find, flying in numbers, a wonderful range of butterflies, from Painted Ladies and Red Admirals to Small Coppers and Small Heaths, with most of the Blues and Browns between, and the occasional Clouded Yellow fresh from its Channel crossing.

It was many years – half a century and more – before I returned to that seaside village, not knowing what I would find, but hoping that the place had not been entirely spoilt. As I followed the coastal path to the village, things began to look very familiar: those rows of small cottages, considerably smarter but little changed, on unmade roads leading towards the pebbly beach. A little farther on was another row of cottages, and across another unmade road, there it was – the enchanted place! That patch of scrubby chalk downland was still there, unspoilt, undeveloped, and now, I was delighted to find, conserved as an SSSI, a site of special scientific interest (or, in my case, special sentimental interest). I found an abundance of orchids, which I don't remember being there in my boyhood. And there were butterflies...

Although the weather was overcast, quite cool, with a fresh breeze and the odd spot of rain falling, the whole space was alive with fresh and frisky Marbled Whites, while Meadow Browns and Ringlets bobbed up from the long grass, and Skippers perched defiantly on their vantage points. As I wandered about in a blissful mnemonic daze, I also spotted Common Blues, Tortoiseshells and Red Admirals... Heaven knows what a wealth of butterflies there must be on a hot and sunny day.

My boyhood butterfly paradise had, amazingly, survived.

It was on that patch of ground that my love of butterflies was born – and that love has remained with me, off and on, ever since. For long periods of my life I could give no more than glancing attention to the butterflies that might cross my path, and had little or no time to go looking for them, but increasingly I would find that, whenever I did have free time and the sun was shining, I would head straight for the nearest countryside (usually the Surrey hills or downs) to see what I could see. Then, when I achieved retirement – the great goal of my working life (as it was for my father) – my love of butterflies came surging back, along with the opportunity to indulge it again, at last. I was now as intensely interested in butterflies as ever I was in my boyhood, and I was finding out more about them than I could ever have known then, thanks to advances in scientific knowledge and optical technology, not to mention good colour printing and the internet (all of which developments have made netting and collecting less and less necessary, even to the serious lepidopterist).

Things came to a head in 2020, the year Covid-19 changed everything. This turned out to be the most intense butterfly year of my adult life, partly because the restrictions of lockdown led me to look closer to home for my butterfly encounters, exploring unfamiliar or overlooked habitats, and being rewarded with a succession of glorious surprises. I became so butterfly-obsessed in the course of that remarkable season that I decided I had better do what I had long been vaguely thinking about: write my butterfly book.

By chronological chance, my own 'butterfly life' illustrates a huge shift in attitudes to the natural world, and in the practices of wildlife lovers: from my boyhood years, when it was still quite normal for amateurs to collect and kill specimens, to the present, when only specialists net butterflies at all, and, for most, to 'catch' a butterfly means to capture its image in a photograph. For myself, I am simply a butterfly *watcher*: I would sooner enjoy the sight of a butterfly with my own eyes than through a lens, though I admire those who can take good photographs of these elusive beauties, so reluctant are most of them to sit still and pose for the camera, and so ready to fly off – to make themselves air, like Macbeth's witches – at the mere glimpse of a lens (I write from bitter experience).

This historical shift from collecting to watching and photographing was driven by changing attitudes, which were themselves a response to an unmistakable loss of abundance in butterfly populations, making conservation the urgent new imperative. Butterflies were noticeably in decline by the 1960s and '70s, and there has been no dramatic reversal of that decline in the decades since. Today, whenever butterflies are mentioned, I find that the first thing people say is that there aren't as many around as there used to be, or even that they never see them. This might be akin to the kind of nostalgia that makes us remember our childhood summers as endlessly sunny, and it surely owes something to the fact that in our younger years we were mostly on foot or on bicycles rather than in cars, and we had more time to look about us. However, the general perception is all too right: there are indeed fewer of them around.

It is hard for us to imagine how numerous butterflies once were in the English countryside. The great butterfly man F.W. Frohawk, writing about collecting in the New Forest in the 1890s, noted that Silver-Washed Fritillaries were then so abundant 'that it was common to see forty or more assembled on the blossoms of a large bramble bush, in company with many White Admirals, Meadow Browns, Ringlets ... When the congregation was disturbed, they would rise in a fluttering mass and the majority would again settle to continue their feast on the sweet blossoms of the bramble.' Another account of collecting in the New Forest in the 1890s talks of swarms 'so thick that I could hardly see ahead, and indeed resembled a fall of brown leaves'.

Michael Salmon (in the preface to his invaluable historical survey *The Aurelian Legacy,* 2000) quotes a collector writing in the 1820s about a trip to Surrey: 'The boundless profusion with which the hedgerows for miles, in the vicinity of Ripley, were enlivened by myriads [of White-Letter Hairstreaks – now one of our most elusive butterflies] that hovered over every flower and bramble blossom last July exceeded anything of the kind I have ever witnessed.' He records, chillingly, that he netted nearly two hundred specimens in half an hour, without moving from the spot. That was as nothing to the carnage wrought by one collector and his friends on the Isle of Wight in the summer of 1875 when they got to work on a swarm of the rare Pale Clouded Yellow and netted some eight hundred specimens. Such wholesale slaughter became commonplace in Victorian times, when astonishing hauls were reported in entomological journals: on a two- or three-day trip to Wiltshire in 1857, for example, one respected entomologist reported netting twenty-eight Silver-Washed, seventeen High Brown and eighty Dark Green Fritillaries, a Large Tortoiseshell, and huge numbers of Ringlets, Speckled Woods, Marbled Whites, Purple Hairstreaks, Chalkhill Blues and Skippers Large and Small. The eighty Dark Green Fritillaries were taken 'in about two hours, without moving beyond a yard or two'.

Such visions of abundance belong to a land very different from the one we know now. They seem to be from a lost land, in which the characteristic patchwork countryside of fields, hedgerows and woods was friendly to man and nature alike, where butterfly-rich downland, grazed by rabbits and cropped by sheep, stretched for uninterrupted miles, woodland was systematically coppiced and harvested for wood and timber, farmland was still rich in wild flowers, and there were no chemical insecticides, artificial fertilisers or traffic fumes. With such an abundance of butterfly life all around them, no wonder most collectors thought it inconceivable that even the rarer species were in any real danger of extinction.

Though their actions were controversial even at the time, some of their number thought nothing of taking out entire colonies of such rarities as the Large Blue and the Large Copper, careless alike of the single life and of the type. Others became obsessed by variants and aberrations, taking vast numbers of specimens of a single species with a natural tendency to variation, such as the Chalkhill Blue. An entire book was devoted to Chalkhill Blue aberrations, co-authored by Percy Bright, sometime mayor of Bournemouth and proud owner of a vast butterfly collection – so vast that, on his death, it had to be sold in five separate auctions, spread over two years. Bright, a classic 'cheque-book collector', habitually toured the countryside in his Rolls-Royce, inspecting other collectors' catches and offering money for anything that caught his fancy. He seems to have relied largely on his chauffeur for the more active side of collecting (and on his co-author, H.A. Leeds, a railwayman and brilliant lepidopterist, for most of the hard work of writing their Chalkhill Blue book).

The routine killing of large numbers of butterflies by those who, ostensibly, most loved them continued well into the twentieth century. Take the troubling case of the Russian-American novelist Vladimir Nabokov, a great butterfly lover and a significant lepidopterologist (that is, one who studies butterflies and moths more scientifically than a mere lepidopterist) who did much valuable research work and named several new species and variants. Nobody wrote more feelingly or more beautifully of the sheer joy of being among butterflies (I shall be quoting him often in the course of this book). And yet Nabokov was a hunter, with a hunter's instincts; his aim was not just to enjoy but to *possess* specimens: to kill them and set them. He would habitually dispatch his smaller victims by pinching the thorax, then place each one in a folded paper slip to keep it fresh until he could further examine it and add it, if required, to his collections. Oscar Wilde's claim that 'each man kills the thing he loves' always struck me as sentimental nonsense, but in a case such as Nabokov's it does seem all too appropriate – and all but impossible to understand from the perspective of our conservation-minded times.

Conservation concerns are now so central to our relationship with butterflies that the main British organisation for butterfly enthusiasts is simply called Butterfly Conservation. This imperative to conserve what we have is felt across the natural world, but is especially acute in relation to butterflies, since all of us, even those who have barely noticed butterflies, are aware that numbers have declined in our lifetime, and that the age of abundance is long over. The figures speak for themselves: the latest comprehensive survey – Butterfly Conservation's *The State of the UK's Butterflies 2022* – shows that eighty per cent of butterflies have declined in abundance or distribution or both since the 1970s, though happily fifty-six per cent have increased in one or both over the same period. On average, the UK's butterflies have lost six per cent of their abundance at monitored sites (though in Scotland abundance has increased by thirty-five per cent) and, more worryingly, they have lost forty-two per cent of their distribution over the period 1976–2019. It is the habitat specialists – butterflies more or less restricted to specific habitats, such as woodland, heathland and flower-rich grassland – that have declined most dramatically, with abundance down by over a quarter and distribution by more than two-thirds. To make matters worse, some once widespread species (like the Graylings and Wall Browns of my boyhood) have become habitat specialists, and therefore much scarcer and more vulnerable. However, there are also former habitat specialists, such as the Speckled Wood and the Comma, that have become all but ubiquitous. And the relatively good news is that non-specialists – butterflies that can adapt to the modern farmed countryside, urban areas and edgelands – have declined far less than the specialists, down less than two per cent in abundance and less than eight per cent in distribution. Apropos these figures, it should be noted that the hot, dry summer of 1976, when this survey began, was marked by unusually high populations of many species, which would have made the subsequent decline seem more dramatic than it really was.

I remember that summer – the driest since 1772, they said – and its extraordinary profusion of butterflies, even in normally unpromising habitats. There has been nothing comparable since, and there was a price to pay the following year, the drought having shrivelled and killed many larval food plants. However, even allowing for the 1976 effect, there is no mistaking the overall downward trend in butterfly numbers, a trend that was already well under way when this longitudinal study began. Then, as if to confirm our worst fears, the results of Butterfly Conservation's 2024 Big Butterfly Count (in which volunteers count how many butterflies they see in a fifteen-minute period on a summer day) came in, showing a dramatic decline in numbers across the board – so dramatic that a 'Butterfly Emergency' was immediately declared. However, it should be borne in mind that the spring and summer of 2024 were relentlessly cool, wet, windy and sunless, so it would have been surprising if the results had been much better.

Why has this happened? Habitat loss is surely the main factor. This has been mostly caused by changes in farming practice: the grubbing up of hedgerows (and harsh machine cutting of those that remain), widespread use of toxic chemicals, 'improvement' of grassland on a huge scale, loss of flower-rich meadows and field margins. Other contributors are a failure to maintain woods and downs, loss of land to urban sprawl, excessive mowing of parks and verges…

The picture is improving in various ways: we are witnessing more relaxed mowing regimes, some restoration of field margins, a little more woodland management, and preservation of patches of butterfly-friendly land. But the butterfly paradise of the pre-war landscape is lost for good, and that wholesale loss of habitat casts a long shadow: the worrying phenomenon of 'extinction debt' means that surviving populations, apparently doing well in odd pockets of the habitat that suits them, are vulnerable to sudden population decline in response to even small changes in their environment. This is one of the reasons why Butterfly Conservation and other organisations are keen to link up patches of good butterfly-friendly habitat into wider swathes of countryside, across which populations can expand and explore.

Butterflies are sometimes spoken of as the 'canaries in the mine' – especially sensitive creatures warning of catastrophe to come – of climate change. However, some higher temperatures are likely to be of benefit to many of our warmth-loving butterflies, provided they are not accompanied by increased rainfall or severe drought. On the other hand, warmer conditions could also benefit butterfly parasites and make some sensitive habitats harder to maintain in the face of more vigorous plant growth. Our hibernating species, too, tend to do better in properly cold winters, which allow them to become fully dormant.

In recent years, there have certainly been more continental species spreading tentatively into the South of England – from the tiny Short-Tailed Blue to the spectacular Continental Swallowtail – and many of our well-established native species have extended their range northward over the years (the Comma, for example, was a prized rarity in Derbyshire in my father's day, but is now to be found all over northern England). These trends are likely to continue. Also on the plus side, many conservation initiatives, especially those aimed at single species, have been successful in reversing decline, the most famous and dramatic example being the reintroduction of the once extinct Large Blue (which I'll be writing about later). However, there is no denying what is going on: a general fall in butterfly numbers which clearly tells us that we must conserve, not collect, these fragile creatures, and that we must value and enjoy them all the more for the precariousness of their situation.

This is how we find ourselves in this late stage of our long relationship with butterflies. How did we get here? How has our enchantment with these beautiful creatures grown and evolved over the centuries? What is the nature of their unique appeal? And what can we gain and even learn from watching them? What, in particular, do they have to offer those in search of a mindful, 'in the moment' way of experiencing nature? Those are some of the questions I shall be exploring in the pages that follow.

Part II

A Flight Through Time

Early Times: A Slow Awakening of Interest

In the early years of the twentieth century, on both sides of the Atlantic, there was a fashion for creating 'Shakespeare gardens', formal gardens designed around plants mentioned in Shakespeare. (Surviving examples include the garden at New Place in Stratford-upon-Avon; and, in America, the Shakespeare gardens in Cleveland, Ohio, the Colorado Shakespeare Garden, and the Shakespeare Garden in New York's Central Park.) There were plenty of Shakespearean plants to choose from: more than fifty species of flowers are named in his works, along with forty-odd trees. Some Shakespeare enthusiasts even tried to introduce 'Shakespeare's birds' into America (which is how New York, and much of the rest of America, got its Starlings and House Sparrows), and again there were plenty to choose from: Shakespeare names over sixty species of bird. He was clearly a man very aware of, and knowledgeable about, the natural world, as well as alive to its beauty. The flower poetry in *A Midsummer Night's Dream*, *Hamlet* and *The Winter's Tale* is some of the most beautiful in the language. And yet there is one notable absence in his works: not a single identifiable butterfly visits his flowers or enlivens his fields and woods. His is a natural world without butterflies; the designers of those Shakespeare gardens had no need to introduce a single species.

The only mentions of butterflies in Shakespeare are purely generic and symbolic, most famously the 'gilded butterflies' that Lear speaks of in the final act of *King Lear*, imagining life in prison with Cordelia: 'So we'll live,/And pray, and sing, and tell old tales, and laugh/At gilded butterflies, and hear poor rogues/ Talk of court news ... ' Clearly he is not thinking of *Lepidoptera*, of actual, living butterflies – why would you laugh at a butterfly anyway? – but of something showy and empty, more akin to the gossiping courtiers Lear talks of in the next phrase. Though it is hard to believe that Shakespeare was wholly unresponsive to the beauty of butterflies, there is no evidence anywhere in his plays or poems that he saw them in terms of individual species, rather than a single generic phenomenon, with some symbolic or metaphorical meaning but nothing more. In this, for all his individual genius, he was a man of his times. Most species of butterfly had simply not been described; they had not yet, in a sense, come into being.

Why was this? Why was it that no one seems to have really noticed butterflies, in all their rich variety and striking beauty? They were surely abundant enough in the unspoilt premodern landscape, and in all probability some species would have had folk names, some of which might have carried over into the names we use today, but there seems to have been insufficient curiosity to distinguish them from each other in any systematic way. But then, why should there have been? What drove the identification of species of plants and larger animals was, above all, utility: the need to know which animals and plants were useful to us as food or medicine, and conversely which could do us harm. Butterflies were among those creatures that had no perceptible human usefulness; neither did they pose a threat, so there was no imperative to know anything about them or to distinguish one from another. That is the practical explanation, but there was another dimension to this incuriosity.

Identification is only one way of fitting living creatures into our human world, of making sense of nature by giving it human meaning. There are other interpretative systems, and the one that held sway before the development of 'science' as we understand it was the all-embracing, richly detailed Christian worldview, in which the whole of nature was seen as God's singular creation. God having created this world and seen that it was good, it must surely be taken to embody His purposes and express His meanings. There were moral, theological and, indeed, practical lessons encoded in nature, and, thanks to the divine gifts of reason and faith, these could be interpreted by man; he could read the book of Nature written for him. An almost literal form of this belief was the remarkably long-lived and pervasive doctrine of 'signatures', which taught that plants resembling various parts of the human body were effective in treating ailments affecting those parts: for example, Eyebright (*Euphrasia*) was widely used to treat eye infections, because of the flowers' perceived resemblance to an eye. How convenient that the Creator had seen fit to make this clear to us.

Right through to the time of the great classifying naturalist Linnaeus and the populariser of 'natural theology' William Paley, and long after the dawn of what we think of as the Enlightenment, it was perfectly possible to see all nature as an expression of the glory and the purposes of God; indeed, it seemed perfectly self-evident. The scientists of the Enlightenment were known as 'natural philosophers', because their business was essentially to investigate nature with a view to interpreting what it was telling us (it being taken for granted that it was there to tell us something). The pre-Darwinian scientific view of nature made an easy fit with the 'natural theology' deployed by Paley to prove the existence of God – and both were to be blown to smithereens by Darwin's shattering idea that Nature had no need of God, and that all of it, man included, was the product of evolution by natural selection. However, old conceptions die hard, and, after the initial explosive impact of *The Origin of Species* (1859), we humans, religious animals that we are, soon found new ways to put ourselves back at the centre of things, to give ourselves and our preoccupations a special and unique significance. Darwin's great insight, like those behind other scientific revolutions, could not be denied, but neither could it be fully internalised (and perhaps that is just as well: it could be argued that the only ones to have really internalised Darwin were the eugenicists whose ideas took hold in the early twentieth century, with terrible results). In our inner sense of being in the world, we carry on as if Darwin, Newton, Einstein, or Galileo never happened. But all that is another story…

For now, let us consider the all-embracing Christian worldview that held sway for many centuries before even natural philosophy was thought of. This comprehensive system incorporated the whole of nature within its interpretative scheme, giving symbolic and dogmatic meaning to as many of God's creatures as seemed worthy. Birds did especially well out of this, with the dove representing the holy spirit, the peacock symbolising immortality, the goldfinch having acquired its red head by pulling a thorn out of Christ's mocking crown, the robin having come by its red breast in shielding the infant Jesus from a fire, and the pelican representing Christ himself (it was believed that the Pelican pierced its own breast with its beak to feed its offspring with its blood). All of these, and more, appear frequently in Christian art, along with a range of other symbolic beasts, flowers and fruits. And what of butterflies? They, it seems, were not deemed worthy of a significant place in this iconography, being seen, in the stern Christian view, as altogether too frivolous, indeed useless; they had no more utility from the spiritual point of view than from the strictly practical. Butterflies make few appearances in medieval imagery and literature, and when they do (typically in drawings in the margins of manuscripts) they seem to function as generic embodiments of silliness, distraction and light-mindedness, flitting unseriously from flower to flower, attracting the attention only of idlers and shallow people. God must have put butterflies on earth as a counterexample, showing us how *not* to be.

However, there was also a much older and deeper-rooted symbolism colouring our perception of butterflies and what they represent – a symbolism that persists to this day. It is evident in the language of the ancient Greeks, who used the same word, *psyche*, for a butterfly and the human soul. The word derives from a demigod, Psyche, who was born mortal and exceptionally beautiful, and granted immortality by Zeus. She was depicted from the first with stylised butterfly wings, emblematic of her liberation from mortality. From there it was but a short step to associating the butterfly in nature with the soul, the immortal part of a person, set free from the temporary housing of the body, as the butterfly emerges from the dead husk of the chrysalis (or appears in this world as if from nowhere, or another world). This association would surely have continued into Christian times, at least at the level of folk belief, but it found no firm place in the Church's iconography. Butterflies scarcely feature on church monuments until the coming of the self-conscious Greek revival in the early nineteenth century, where clearly they carry much the same symbolic meaning as in pre-Christian times, while also representing the fleetingness of life. Though butterflies were marginalised in the symbolic worldview that underpinned Christian doctrine, the association with the liberated soul that was born in much earlier times proved stronger and deeper than anything in that system – so strong indeed that it has come surging back in our own times. I shall be writing more of that extraordinary persistence later.

Whatever the prevailing worldview, it is hard to believe that, in pre-Christian and Christian times alike, people did not look at the butterflies around them and experience, if not scientific curiosity or an urge to classify and describe, then at least some sense of the beauty and mystery of what they were looking at, some dawning enchantment. How could they not have wondered, at least, where these creatures had come from when they suddenly appeared every spring and summer? Birds had hatched from eggs, mammals had been born, but butterflies, like less attractive kinds of fly, must have seemed to have generated spontaneously, or by some quite unknown process. By Shakespeare's time, a kind of protoscientific curiosity was beginning to stir in scholarly circles, and some tentative attempts to distinguish and describe various species of British butterfly were under way. The first fruit of this early endeavour was a volume grandly titled *Insectorum sive Minimorum Animalium Theatrum* (*Theatre of Insects*) (1634), edited and, according to him, greatly improved by one Thomas Moffet (or Muffet, Moffat or Mouffet), a physician who had studied silkworms in Italy, and was particularly fascinated by spiders. A conflation of the insect studies of the Swiss polymath Conrad Gesner and two English 'students of nature', Edward Wotton and Thomas Penny, the *Theatre* had a long and rocky road to publication, much of the voluminous source material having at one point been torn into pieces by Penny's niece and only rescued in the nick of time by Moffet. Then, having laboriously wrestled the book into shape, cutting out 'a thousand tautologies, trivial matters and things unseasonably spoken', Moffet could find no one willing to publish it; there was no great demand for books of natural history, especially if written in Latin. Eventually, in 1634, thirty years after Moffet's death, the *Theatrum* saw the light of day, published in a form considerably reduced from the original grand design and with inferior woodcut illustrations. It became more popular after it was translated into English and included as an appendix in Edward Topsell's *History of Four-footed Beasts and Serpents* (1658). No better book on insects was to be had until John Ray (see below) got to work.

As for the butterflies in Moffet's *Theatrum*, they are not named, only described in Latin. From these descriptions, some eighteen British species have been identified with varying certainty, along with the continental Apollo and Scarce Swallowtail. It was to be a long time – some eighty years – before a substantial number of British butterflies were named, in English, in a book. When the great all-round naturalist John Ray turned his attention to butterflies, sadly late in his career, at a time when he was afflicted by illness and living in straitened circumstances, he too had no names to give to individual species, and could only describe them in Latin. His *Historia Insectorum*, incomplete at the time of his death, was edited and published posthumously in 1710, and was a considerable advance on anything that had gone before, describing forty-eight species (along with some three hundred moths), and describing them accurately enough for them to be identified today, at least by those capable of deciphering the Latin. Ray's contribution to natural history was huge, a body of work built on close observation and accurate description, and including a rational system of classification that would not be improved on until the great Swedish naturalist Linnaeus got to work. Of this great enterprise, his study of butterflies was only a minor part – and yet Ray left one of the most eloquent justifications for taking an interest in these seemingly useless creatures. It was written originally in Latin, of course, but in English it reads: 'You ask what is the use of butterflies. I reply, to adorn the world and delight the eyes of men: to brighten the countryside like so many golden jewels. To contemplate their exquisite beauty and variety is to experience the truest pleasure. To gaze inquiringly at such elegance of colour and form designed by the ingenuity of nature and painted by her artist's pencil is to acknowledge and adore the imprint of the art of God.' Here, for the first time, is the true language of enchantment.

Even after John Ray's endeavours, the study of butterflies was still the preserve of those who were at home in the Latin language and had the leisure to pursue an interest in natural history. Before butterflies could become more general objects of interest they needed one thing above all – English names. The first man to systematically name our butterflies in our own language was James Petiver, a strange, obsessive man whose passion in life was his personal museum, a huge, disorganised collection of natural history specimens of all kinds. Making use of a worldwide network of contacts, he amassed material from far and wide, and was none too scrupulous in his pursuit of funds to support his endeavours. Along the way he published papers and pamphlets on a range of subjects, before finally – as with Ray, near the end of his life – amalgamating his notes on British butterflies into a single slim volume, *Papilionum Britanniae Icones* (1717), which contained images of and basic information about 'Eighty British Butterflies' and gave them all English names, some of them his own inventions, others carried over from folk names. Even in Petiver's time there were not as many as eighty British species; some of his supposedly distinct species are in fact males and females, or variants of the same butterfly, and one that he includes, 'Albin's Hampstead Eye', is now known to be an exotic species, occurring only on certain islands in the Pacific and Indian oceans (the origins of Petiver's Hampstead specimen are a mystery). Many of the names given by Petiver – Admirals, Hairstreaks, Arguses, Tortoiseshells – are still in use today, while others, such as 'Hogs' for Skippers and, bizarrely, 'Royal Williams' for Swallowtails, did not endure. The *Papilionum* is a slight piece of work, but it was the first time anyone had published anything like a comprehensive and systematically classified list of British butterflies and given them all English names. It was enough for him to be thought of ever after as 'the father of English butterflies'.

A Slow Awakening of Interest

Despite his unattractive personality and apparently chaotic working methods, Petiver was a brilliant 'networker', as we would say today, building a small army of contacts – fellow naturalists, collectors and travellers – through whom he extended his knowledge and expanded his vast collections with specimens from all corners of the world. Closer to home and on a smaller scale, networks of butterfly enthusiasts were, by the early years of the eighteenth century, increasingly active, sharing knowledge and swapping specimens. One naturalist who was ubiquitous in these circles was Joseph Dandridge, who, had he been less self-effacing and keener to publish, might himself have become the 'father of English butterflies'. He was spoken of by his peers with great respect, but he seems to have been happy simply to help his fellow collectors and share his knowledge and enthusiasm, rather than seek a more lasting fame by publication. By trade he was a pattern designer for silk weavers, and he drew and painted the objects of his interest, but his pictures were never published, and most were lost. The story is told of him that on one occasion a farm labourer, seeing him wildly lunging at the air, took him for a lunatic and wrestled him to the ground to subdue him. He was, it seems, about to net a Purple Emperor: 'The Purple Emperor's gone!' he wailed. 'The Purple Emperor's gone!' – which only served to confirm the labourer's initial impression.

Dandridge was an early victim of a popular tendency to equate lepidoptery with lunacy, and view with deep suspicion anyone wielding a butterfly net.

Even at the height of the Victorian butterfly collecting mania, the eminent naturalist Edward Newman wrote that 'ninety-nine persons out of a hundred' were of the opinion that anyone 'who could take an interest in pursuing a butterfly is a madman'. Or a madwoman …

Eleanor Glanville, another early collector and one of the few women in an overwhelmingly male field, was to fall foul of this suspicion in a much more serious way. Glanville (sometimes referred to as Lady Glanville, though she was not in fact titled) took up lepidoptery in middle age, after the breakdown of her marriage to her second husband, a brutish Lincolnshire landowner, Richard Glanville, who on at least one occasion held a loaded and cocked pistol to her breast and threatened to shoot her dead. Following the marital breakdown, the unscrupulous Glanville was determined to get his hands on Eleanor's money – she was a woman of property – by hook or by crook, on one occasion kidnapping one of her sons in the hope of getting him to disclaim his inheritance and transfer it to him and his new mistress. This kind of thing led Eleanor to arrange to leave the disposal of her estate in the hands of trustees. However, on her death in 1709, the will was disputed by her eldest son, whom Richard Glanville had turned against her, on the grounds that his mother had clearly gone mad. One of the signs, sure enough, was her pursuit of butterflies. As well as beating bushes to knock 'worms' (larvae) out, she had been seen chasing around all over the countryside, often 'without all necessary cloathes' and even dressed 'like a gypsey'. This kind of behaviour was suggestive not only of lunacy but also, in an age that still believed in such things, witchcraft. In the end, the will was indeed upset, and the eldest son inherited. It was agreed by one and all that no one 'not deprived of their sense should go in pursuit of butterflyes'. After Eleanor Glanville's remarkable life story was pieced together by scholars in the 1960s, it attracted a good deal of interest, and even inspired a romantic novel, *Lady of the Butterflies* by Fiona Mountain (2009). She is surely the only lepidopterist, male or female, to have been so honoured.

Eleanor Glanville, far from being a lunatic, was a considerable lepidopterist, some of whose specimens are among the earliest preserved in the Natural History Museum. She even has the almost unique distinction of having a British butterfly named for her – the lovely Glanville Fritillary, which she discovered in Lincolnshire, but which is now limited to a number of sites on the Isle of Wight (and a few introductions elsewhere). She corresponded on equal terms with Petiver, Ray, Dandridge and other collectors, and sent them large numbers of specimens. When she died, in 1709, the world of British butterfly enthusiasts was a small but increasingly lively one, with new discoveries being made all the time, and much swapping of specimens and naming and classification of species going on. From this ferment, a product of the new-found scientific interest in the natural world that welled up in the post-Restoration years, the first organised entomological society was to be born, and the British love of butterflies was to enter a new, vivid and rather delightful phase – the age of the Aurelians.

The Eighteenth Century: Enchanted Aurelians

It is easy to talk blithely of a 'new-found scientific interest in the natural world' in the post-Restoration years – indeed, I have just done so – and to file this development under 'the Enlightenment', the intellectual movement that supposedly saw the light of Reason scattering the obscuring clouds of superstition and ignorance. However, as is usual in history, things were not that simple. The background against which the new interest in butterflies developed was more like a complex tapestry than a clearly delineated picture – a tapestry in which many different strands, some recognisably scientific, others more imaginative and spiritual, were interwoven. What is clear is that the early lepidopterists, the 'aurelians' as they styled themselves, were driven at least as much by aesthetic delight (and a religious sense of the wonders of creation) as by the spirit of pure scientific inquiry, and their great works – the magnificent illustrated books of the eighteenth century – were, as we shall see, a perfect and beautiful expression of both.

Science in the late seventeenth century, even as practised by the luminaries of the Royal Society (founded 1661), was in many ways a very different beast from science as it is now. For one thing, it generally took for granted the role of God, or a divine being, as the prime mover and creative force behind any scientific model of the universe and its workings – a distant God, perhaps, who had little to do but wind up the clockwork and sit back, but a real and necessary divine being nonetheless. John Ray, a devout Christian, might have published much ground-breaking work in the field of natural history, but his most popular book was the *The Wisdom of God Manifested in the Works of the Creation* (1691), which ran to seven editions. Isaac Newton, the supreme scientist of his age, whose model of the universe held sway for two centuries, was obsessed by theological questions, wrote extensively on them, and firmly believed in an omnipotent God. 'This most beautiful system of the sun, planets and comets,' he writes in the *Principia* (1687), 'could only proceed from the counsel and dominion of an intelligent and powerful Being ... This Being governs all things, not as the soul of the world, but as Lord over all.' The idea that science was a self-sufficient system that could ultimately explain everything, without recourse to anything resembling Newton's all-designing Being, would never have occurred to the 'natural philosophers' of Newton's time. He himself, having come close to creating something very like a Theory of Everything, famously concluded that 'to myself I seem to have been only like a boy playing by the sea shore, and diverting himself in now and then finding a smoother pebble or a prettier shell than ordinary, whilst the great ocean of truth lay all undiscovered before me'. Compare and contrast this humility with Stephen Hawking's famous assertion that 'if we do discover a theory of everything ... then we would truly know the mind of God'. Would we indeed? Ludwig Wittgenstein, perhaps the greatest philosopher of the twentieth century, took a more measured view: 'Even if all possible scientific questions be answered, the problems of life have still not been touched at all. Of course there is then no question left, and just this is the answer.'

Newton was also fascinated by alchemy, to such an extent that something like a tenth of his surviving writings are devoted to that arcane – and to modern eyes, futile – field of inquiry. Chemistry had yet to be fully separated from alchemy, and, more generally, the borders between science and religion, and between science and something more like magic, were quite porous at this time. One of those who tried to distinguish between the realms of science and popular superstition was Sir Thomas Browne, a writer of wonderfully musical prose, whose most voluminous (and, in his day, most popular) work was *Pseudodoxia Epidemica, or Enquiries into very many received tenents and commonly presumed truths* (1646). This employs the empirical method pioneered by Francis Bacon to separate fact from fancy, but at the same time draws heavily on the authority of past authors, especially the ancients, whose wisdom of course predated anything we would recognise as 'science'. Scientific inquiry, for Browne, could only take him – or anyone – so far, and he liked it that way: 'I love to lose myself in a mystery, to pursue my reason to an *O altitudo*.'

Similarly, the borders between science and the arts, between scientific inquiry and a lively sense of beauty and wonder, were fluid, and remained so for a long while. When, in the last years of the eighteenth century, the polymath Erasmus Darwin, grandfather of Charles and proponent of an early theory of evolution, published his writings on a vast field of natural history, he did so in handsome volumes of florid verse, full of fanciful imagery, drawing heavily on classical mythology and the Rosicrucian system that populated the natural world with sylphs, nymphs, gnomes and salamanders. Here, as an example of his overheated style, is a passage from *The Temple of Nature* (one which deservedly found its way into that great anthology of bad poetry, *The Stuffed Owl*, of 1930):

> So still the Tadpole cleaves the watery vale,
> With balanc'd fins and undulating tail;
> New lungs and limbs proclaim his second birth,
> Breathe the dry air, and bound upon the earth.
> Allied to fish, the Lizard cleaves the flood
> With one-celled heart, and dark frigescent blood;
> Half-reasoning Beavers long-unbreathing dart
> Through Eirie's waves with perforated heart;
> With gills and lungs respiring Lampreys steer,
> Kiss the rude rocks, and suck till they adhere;
> With gills pulmonic breathes th' enormous Whale,
> And spouts aquatic columns to the gale.'

This sort of thing seems a far cry from science as we know it, but one of Erasmus Darwin's works, *The Loves of the Plants* (1791), was an exposition in verse of the key findings of Linnaeus, and, in the course of his poetic effusions, Darwin did expound the idea that all forms of animal life might have a common ancestor. At around the same time as Darwin, the Swedish naturalist's insights into botanical science also inspired one of the most beautiful of all illustrated books, Robert Thornton's three-part *New Illustration of the Sexual System of Carolus Linnaeus*, culminating in *The Temple of Flora* (1799–1807). But, to return to our theme, and to an earlier phase of the English scientific enlightenment, the possibilities of the illustrated book were already becoming apparent in the late seventeenth century. They were never more dramatically demonstrated than by Robert Hooke's *Micrographia* (1665), with its gigantic, startling images of insects seen through the microscope. These are beautiful images in themselves, and unmistakably suffused with a sense of wonder at the new worlds opened up by scientific investigation. As for the lavishly illustrated butterfly books that began to appear in the early eighteenth century ... That is where we are heading now, to the colourful, companionable world of the Aurelians.

The early eighteenth century was the golden age of the coffee house, that convivial space where men – and it was almost exclusively men – would come together to drink coffee, smoke, read, transact business and exchange gossip, news and views. London was said to have more coffee houses at this time than any city in the world, other than Constantinople. Customers with common interests tended to gravitate towards particular coffee houses, or taverns of the more respectable kind, where informal associations of the like-minded grew up. In the City of London, entomologists (the word 'lepidopterist' was not yet in use, and besides, butterflies and moths were not the only insects these enthusiasts were interested in) gravitated not to a coffee house, but to the Swan Tavern in Exchange Alley, a venue well known for concert performances. Was this part of the attraction? It might well have been, for these early butterfly fanciers seem to have been artistically inclined and certainly alive to beauty. At some time between 1720 and 1742 they decided to formalise their association, thereby establishing the first entomological society in history, and they gave themselves a splendid name – the Society of Aurelians. The word 'aurelian' is derived from the Latin ***aureus***, meaning golden. This probably refers to the gold spots that adorn the chrysalids of certain (Nymphalid) butterflies – indeed the word chrysalis derives from the Greek ***khrusos***, meaning gold – but it might have been chosen simply for its Latinate lustre and its glamorously suggestive sound. 'Aurelian' is certainly preferable to any later name for butterfly lovers: 'lepidopterist' is clumsy and sounds like a medical specialist ('I'm going to refer you to a lepidopterist'). Some, notably Vladimir Nabokov, have done their best to keep the word 'aurelian' alive. It certainly fitted those early butterfly men (and women) perfectly.

I confined the women between parentheses there because the Society of Aurelians itself had no women members. However, as we have seen from the example of Eleanor Glanville (see previous chapter), there were some very active women in the butterfly world of the time. Among them were society ladies, with whom entomology was a fashionable interest (generally more aesthetic than scientific) even before the Society of Aurelians was formed. Lady Margaret Cavendish Bentinck, Duchess of Portland, had the largest and most famous natural history collection in the country, a compendious museum, including a wide range of butterflies and moths, housed at Bulstrode Park in Buckinghamshire. A woman of intellect, one of the famous Bluestockings, her ambition was 'to have had every unknown species in the three kingdoms of Nature described and published to the world'. Her interest in nature was seriously scientific as well as aesthetic; and the same could be said of another aristocratic naturalist, Mary Somerset, Duchess of Beaufort, a distinguished gardener and botanist, whose twelve-volume herbarium, bequeathed to Sir Hans Sloane, is in the Natural History Museum. She is known to have reared butterflies and moths herself, and was a patron of the entomological artist Eleazar Albin (of whom more soon). Well into the eighteenth century, society ladies continued to take an interest in Lepidoptera; it was no mere passing fashion. A quarter of the subscribers to Benjamin Wilkes's sumptuous 1750 volume *English Moths and Butterflies* were women, as were a third of the dedicatees of Moses Harris's plates for *The Aurelian*, published in 1766 (more on both of those books later).

The Society of Aurelians, however, continued as an all-male preserve throughout its life. Little is known about the society's activities because its records and collections, library and regalia were all lost in a fire that swept dramatically through Exchange Alley on a March night in 1748, as reported at the time by the *Entomologist's Weekly Intelligencer*, forcing the Aurelians themselves to flee for their lives: 'So sudden and rapid was the impetuous Course of the Fire, that the Flames beat against the Windows, before they could well get out of the Room, many of them leaving their Hats and Canes.' Sadly, the effects of the fire so disheartened the members that the society was in abeyance for fourteen years before it was briefly resurrected, only to fail again. A third Society of Aurelians, founded as late as 1801, was also short-lived: the Aurelian moment had passed, and entomology was by then becoming a more solidly scientific affair, with rather less of the gaiety and *brio* of the early days.

What survives of that first Society of Aurelians is a general impression of a group of genial, fun-loving, rather dandyish men with a lively interest in butterflies. The society's leading lights – including its probable founder, Joseph Dandridge (see previous chapter) – were artists, illustrators and textile designers as well as aurelians, and we know them for the remarkable books they produced, products of an age when, to quote the social historian G.M. Trevelyan, 'life and art were still human, not mechanical, and quality still counted far more than quantity'. Before we look at those, we need to take a small detour to pay tribute to the brilliant artist and entomologist who paved the way for all those who came after – an artist who was neither English nor a man. The intrepid Maria Sibylla Merian, born in Frankfurt in 1647 into a family of artists, took an early interest in silkworms, which inspired her to study the process of metamorphosis, and in time to fill two volumes with her own expertly engraved and etched plates of caterpillars. At the turn of the eighteenth century, she spent two years in Suriname, a then Dutch colony in South America, studying and drawing the plant and insect life, and taking a particular interest in the different life stages of butterflies and moths. She was arguably the first to see and illustrate the life of the butterfly as a whole – egg, caterpillar, chrysalis, butterfly, food plants and habitat, the complete ecological package. Her *magnum opus*, the *Metamorphosis or Transformations of the Insects of Surinam* (1705), contains sixty sumptuous plates, each of them a little masterpiece. They are at once scrupulously accurate and informative illustrations, drawn from life, and exuberant, sinuous works of art, bursting with life, carrying the great Dutch tradition of flower (and butterfly) painting into, literally, a new world, and mingling it with the decorative freedom and elegance of the Rococo. Nothing produced in England in the eighteenth century quite matched Merian's bravura productions, but the best of it sometimes came very close.

The first to publish a volume containing colour plates of British butterflies was Eleazar Albin (who gave his name to that mysterious specimen, 'Albin's Hampstead Eye' – see previous chapter). Little is known of his origins: he might have been German-born, he might have spent time in Jamaica. He was a painter by profession, and his interest in insects was closely related to his artistic activities: as he recalled, 'teaching drawing and Paint in Water-Colours being my Profession, first led me to the observing of

Flowers and Insects, with whose various Forms and beautiful Colours I was very much delighted.' Among those he taught were his own daughters, whom he trained to assist him in producing his insect illustrations. As mentioned above, he had the patronage of Mary, Duchess of Beaufort, and that enabled him to build a list of well-to-do subscribers for an illustrated book on insects. This was *A Natural History of Insects*, published in 1720, a volume containing a hundred copper plates, only fifteen of them portraying butterflies (most of the rest are moths). Despite its price tag – three guineas coloured (the equivalent of nearly four hundred pounds today), thirty shillings plain – it was a success, running to five editions. Like Maria Sybilla Merian, Albin, who is known to have reared butterflies himself, typically showed the adult insect with the early life stages and the larval food plant. His *Natural History* was the first book to portray British butterflies in more or less lifelike poses and in full colour; the plates, while lacking the vivid tones, vitality and compositional flair of Merian's, set a new standard for entomological art in England.

Benjamin Wilkes produced something a good deal livelier with his *English Moths and Butterflies* (1749), a volume full of very attractive coloured plates depicting butterflies and moths and plants, brought together into pleasing, decidedly artistic compositions. In some plates various stages of the life cycle are shown, while in others a number of different species of butterflies and moths, wings spread as if on the setting board, are arranged into purely decorative geometric compositions (a style of presentation that would become popular with the Victorians, not to mention that master of kitsch Damien Hirst in his 'butterfly' phase). Wilkes was a painter of history pieces and portraits who became fascinated by butterflies and moths after attending a meeting of the Society of Aurelians; it is noted in the preface to *English Moths and Butterflies* that 'here he first saw such Specimens of Nature's admirable Skill in the Disposition, Arrangement and contrasting of Colours ... as struck him with Amazement, and convinced him, at the same Time, that studying them would turn greatly to his Advantage.' Devoting his spare time to collecting, observing and drawing them in all their life phases, he was able to produce in 1742 *Twelve New Designs of English Butterflies*, a dozen engraved plates of butterflies and moths, with some information about each. Its splendid frontispiece contains a beautifully lettered dedication 'to the worthy members of the Aurelian Society' in a cartouche surrounded by a picturesque arrangement of foliage and caterpillars. His *English Moths and Butterflies* (1750) included many species illustrated for the first time, among them two of our most spectacular moths, the Clifden Nonpareil and the Kentish Glory, and the lovely Camberwell Beauty, identified baldly as the 'Willow-Butterfly'. The book was a commercial success, and new editions were published in 1773 and 1824, the last incorporating Linnaean nomenclature. Wilkes himself did not long outlive his book, dying suddenly soon after its completion.

The most charming, accomplished and beautiful of the Georgian butterfly books is Moses Harris's *The Aurelian* (1766), which again brings butterflies and moths together with plants, in plates that are at least as attractive as Wilkes's, while being a good deal more artistically composed. Little is known about Harris – even the date of his death – but he described himself (in *The Aurelian*) as a 'Painter, who has made this Part of Natural History his Study and has bred most of the Flies [butterflies] and Insects for these twenty years'. The forty-one plates in the first edition of *The Aurelian* show some thirty-nine butterflies (with ninety-three moths, four beetles and a dragonfly), with their early life stages carefully depicted. The butterflies and moths, clearly drawn from living specimens, are not only accurate but full of life, and each plate brings several species together into a complete, balanced artistic composition, often incorporating a vase of flowers or some other not entirely relevant objects, included to perfect the finished plate. *The Aurelian* is also famous for its delightful frontispiece, depicting Harris himself (almost certainly), seated by a woodland ride, elegantly dressed and posing at his ease, his long, two-handled net over his knees, some of his prized catches displayed around him in oval boxes. The setting is gloriously sylvan, and in the ride behind him a fellow enthusiast stalks his prey. It is an enchanted scene, an English aurelian's vision of paradise, and beneath it are inscribed words from Psalm 111: 'The works of the Lord are great, sought out of all them that have pleasure therein.'

There is one more Georgian butterfly book that must be mentioned – the gorgeous six-volume work known as 'Jones' Icones'. The handiwork of a wealthy wine merchant and amateur entomologist called William Jones, it consists of some one thousand five hundred beautiful, meticulously painted watercolours of butterflies and moths from around the world, some of them brought back by Joseph Banks from Captain Cook's first expedition, many others painted from specimens in the collections of Jones's collector friends, including Dru Drury, who himself produced a handsome volume of illustrations of exotic insects, *Illustrations of Natural History* (1770). Unlike Drury's volume, Jones' Icones were never published, and the original manuscript resides in the Oxford Museum of Natural History. It was recently digitised and can be studied online.

The Aurelians represented what might be termed the artistic or aesthetic phase of lepidoptery, a phase that was to end with the rise of a more strictly scientific study of butterflies, based on systematic classification. As noted above, by the time of the third edition of Benjamin Wilkes's *English Moths and Butterflies* (1824), Linnaean binomials were standard. Before that, the failed attempt to relaunch the Aurelian Society would seem to suggest that as early as 1801 the Aurelians had had their day. The Linnaean Society, founded in 1788 with the purpose of 'cultivating Natural History in all its branches', was now offering a new model of learned society for the study of natural history, one that was to give rise to a number of new organisations, including the Entomological Society of London, established in 1833. Although the inaugural meeting took place in a tavern – the Thatched House in St James's Street – this was a more sober, organised and rigorously scientific institution than the Aurelian Society. Times had changed.

By the early years of the nineteenth century, the groundwork for the study of butterflies had been laid by the early pioneers, by the artistic Aurelians, by the classifiers and namers, and the members of the new learned societies. What was to happen next was an explosion of interest among a much wider public, leading to the establishment of large numbers of field clubs and natural history societies across the country, a profusion of books and journals – and a golden age of collecting, at every level from the schoolboy hobbyist to the scientifically minded lepidopterist. This might have added greatly to the gaiety of the nation, but it was not, ultimately, good news for our butterflies. It is that 'golden age' that I shall be exploring in the next chapter.

The Victorian Age: Men with Nets

For the Aurelians, enlightenment was closely interwoven with enchantment: the beautiful books that are their legacy are clearly products of enchantment as much as scientific curiosity. And there is something very British about this. For all their reputation for hard-headed empiricism, the British also have a weakness for enchantment, mystique and fantasy, as demonstrated by our long-standing prominence in the field of fantasy literature, from Lewis Carroll to J.K. Rowling by way of Tolkien, C.S. Lewis and many others. Until very recently, the British have liked to clothe even mundane realities with an air of fantasy and bogus antiquity. Consider the English facility for inventing 'traditions' and making them seem as if they've been around for ever: harvest festival, morris dancing, the nine lessons and carols at Christmas, Remembrance ceremonial, much of the pageantry of Parliament and Crown. Consider that most fantastically faux antique style of architecture and design, the Gothic revival; only in the English-speaking world did it sweep all before it, lending a magical air of history and myth to everything from mansions and palaces to new-built churches and chapels, town halls, suburban villas, even office buildings and factories. Amid the dark satanic mills, the Victorian British could not resist the urge to lend enchantment to every view, and the flavour of an imagined medieval past to all the externals of life. This nostalgia for a past that never was found spectacular (and spectacularly silly) expression in the Eglinton Tournament of 1839 in which, over three days, an assembly of aristocrats, suitably arrayed, re-enacted a full-dress medieval joust and revel at Eglinton castle in Ayrshire. In church architecture, the Gothic revival amounted to, in Roger Scruton's phrase, 'an evangelism of enchantment', and the British proved uniquely receptive to it; the desire for enchantment, the willingness to fall under a spell, was never far below the surface.

The Aurelians also represented another marked strain of the British: the urge, particularly among men, to join together in clubs, fellowships and societies of various degrees of formality, pursuing common interests in a sociable, often highly convivial manner. 'Field clubs', many of them associated with the literary, philosophical and scientific societies that burgeoned in the nineteenth century, were popular and characteristic features of the study of natural history in Britain. These would invariably lay down an impressive formal framework of meetings, minutes, regalia, collections and proceedings, while at the same time leaving plenty of scope for fun. Picnic outings, open to both men and women, were especially popular, offering plenty of food and drink, and often music and dancing, as well as a little open-air instruction. The presence of ladies on these occasions was not to everyone's taste. When Charles Kingsley, the famous author (of *The Water Babies* and much else) and muscular Christian, came to Chester as Canon of the cathedral, he founded the Chester Society of Natural Science, Literature and Art. This proved to be a runaway success, and the Society's excursions attracted many wives, daughters and girlfriends, much to the disgust of Kingsley, who was once heard to remark: 'Those good ladies quite spoilt my day – but what *can* you do? When they get to a certain age you must either treat them like duchesses or sh-sh-shoot them!' (He had a slight stammer that he never quite shook off.) However, there was no way of keeping the ladies away from the fun, and the field clubs' grand social occasions continued to offer a wonderful blend of edification and entertainment, lavish hospitality and good cheer.

In the Victorian period, the butterfly fancy ceased to be a niche pursuit, a London-based phenomenon that was the preserve of a few relatively well-off enthusiasts and artists; it became, in a word, democratised, with entomological clubs forming in all parts of the country, and the pursuit of butterflies and other insects becoming popular at all levels of society. In her novel *Mary Barton* (1848), Elizabeth Gaskell paints a fascinating picture of the range of interests of hand-loom weavers in the north of England in the 1840s, men who take 'real scientific delight' in catching 'any winged insect', surely including butterflies:

> *In the neighbourhood of Oldham there are weavers, common hand-loom weavers, who throw the shuttle with unceasing sound, though Newton's Principia lie open on the loom, to be snatched at in work hours, but revelled over in meal times, or at night. Mathematical problems are received with interest, and studied with absorbing attention by many a broad-spoken, common-looking factory hand. It is perhaps less astonishing that the more popularly interesting branches of natural history have their warm and devoted followers among this class. There are botanists among them, equally familiar with either the Linnæan or the Natural system, who know the name and habitat of every*

> *plant within a day's walk from their dwellings; who steal the holiday of a day or two when any particular plant should be in flower, and tying up their simple food in their pocket-handkerchiefs, set off with single purpose to fetch home the humble-looking weed. There are entomologists, who may be seen with a rude-looking net, ready to catch any winged insect, or a kind of dredge, with which they rake the green and slimy pools; practical, shrewd, hard-working men, who pore over every new specimen with real scientific delight.*

The keen interest in Newton's *Principia* might not have been quite so widespread among the working classes as Mrs Gaskell suggests, but an enthusiasm for 'the more popularly interesting branches of natural history' was by this time widely diffused through society, from high to low, and flavoured the way natural history was studied in this country with direct, close observation in the field, rather than from theoretical models and ideal forms. The latter approach, ultimately derived from Platonic philosophy (the British were always more Aristotelian), prevailed in continental Europe, especially France and Germany (and, incidentally, fuelled a strong resistance to Darwinian evolutionary theory in those countries). In empirically minded Britain, field work drove the process of discovery, involving specialists and scholars, but also an ever-widening range of people of all sorts and conditions: from those quintessentially English figures, the parson-naturalist and the naturalist squire, through the various levels of the middle class to clerical and industrial workers, railwaymen and factory hands. It is worth recalling that, even in the grimmest of northern industrial towns, workers were unlikely to be more than a fairly short walk from the countryside, and that many of these workers still had their roots firmly in the land.

Various factors facilitated the spread of a practical interest in natural history – butterflies very much included – through society in the nineteenth century: a general rise in living standards and increase in leisure time, despite huge inequalities of wealth; the spread of the railways, which, along with improved roads, dramatically transformed the ease and speed of travel around the country; the coming of a cheap and reliable postal service with the penny post; a little later, the development of the bicycle, opening the countryside to all; cheaper printing and paper, enabling the production of low-cost books and magazines; and the production of cheaper, easier to use equipment for the collector (of which more in the next chapter). From the mid-nineteenth century on, butterfly collecting became ever more popular, reaching a peak in the Edwardian period, and continuing to be keenly pursued by many (my father included) through the interwar years.

The Victorian butterfly world was suffused with the unbounded energy, enthusiasm and hearty self-confidence of the age, and with an equally characteristic strain of muscular, self-improving morality that would have been quite alien to the more hedonistic Aurelians.

In his *Lepidopterist's Guide*, a popular primer published in 1869, Dr H.G. Knaggs lists among the 'Objects for which Men Become Entomologists', 'Pursuit of truth, with a love of nature, and a laudable desire to investigate the histories of the wonderful organisms which God has, in his wisdom, created.

> *Healthful occupation for those who have spare time to devote to a pursuit which leads them, with an object constantly in view, to green fields, country lanes, sunny banks, shady groves, noble parks, and makes them familiar with beautiful scenery. Emulation – a desire to excel, an ambition to possess the finest collection, to be considered the best collector, to be known as a most accurate observer, or to be handed down to posterity as a great nomenclator.*
>
> *Acquisitiveness, the feeling which actuates the schoolboy to hoard up marbles, buttons ... birds' eggs, and postage stamps. It is, at any rate, better to gratify the propensity (when we are unfortunate enough to possess it) by collecting, than it is to become a wretched miser.*
>
> *The good effects of Entomology are numerous: patience, perseverance, and punctuality are essential for successful collecting; memory, discrimination, and logical reasoning are necessarily cultivated; early rising is encouraged; the mind and body of youth find occupation; temptation to immoral pursuits loses its effect; and liberality with a desire to assist brother collectors is generally engendered ...*

Dr Henry Guard Knaggs, a north London GP with a keen interest in lepidoptery, embodies both the moralising strain and the cheerful sociability of the Victorian collecting world. His house in Camden Town, still more or less a village in those days, was a popular meeting place for entomologists attracted by Knaggs's generous and hospitable nature. After meetings of the Entomological Society, Fellows would often reassemble at Dr Knaggs's house for refreshment and good cheer. This conviviality sat easily with a sense of moral, social and educational purpose; both served to spread the gospel of natural history, the 'evangelism of enchantment' in another form. The wealthy and well-connected Henry Tibbats Stainton was another entomologist who was conscious of the moral value of his pursuit and keen to encourage it, not least by holding regular 'at homes' in his Lewisham house, to which all were welcome, from beginner to expert. These popular 'at homes' ran parallel to those of another collector, J.F. Stephens, and Stainton's continued after Stephens's death. The purpose of such 'at homes' was to enable collectors to swap expertise; to sort out knotty points of entomology; to view the host's collections; and, of course, to chat, gossip, eat and drink, and have a jolly time.

Another who kept open house, in his case every Thursday evening, was Edward Newman – and here we meet one of the greatest, and most attractive, figures of the Victorian butterfly world. Newman, luxuriantly bearded in the Victorian manner, looks good-humoured and relaxed even in photographs, reflecting his open, genial personality. It is surprising (though perhaps it shouldn't be) to learn that he was prone to short-lived fits of what we would now call depression. In 1837, while still in his mid-thirties, he wrote, 'I think that the opportunity for enjoying life with me will shortly expire, and I am desirous, while blest with strength and health, of visiting the country ...'. In the event, Newman lived into his mid-seventies, in good health until his final years. When he wrote the passage quoted, he was weighed down with business concerns, as his early ventures into commerce had all gone nowhere, and he faced an uncertain future. However, a few years later, he got married and became a partner in what was, for him, the perfect business: a firm of London printers, Luxford & Co, which, with Newman on board, became a leading publisher of books on natural history, including Newman's own. From then on, there was no stopping Edward Newman, whose first publication, *A History of British Ferns* (1840), soon became a classic. A formidably gifted all-round naturalist, whose interests ranged far beyond butterflies and moths, he became editor of *The Entomologist* and *The Zoologist*, and natural history editor of *The Field*, and published, among much else, two important books on Lepidoptera: his *Illustrated Natural History of British Moths* (1869) and *Illustrated Natural History of British Butterflies* (1869).

While still in his early thirties, Newman was elected editor of *The Entomological Magazine* (the journal of the Entomological Club, of which he was a founder member), and became a fellow of the Linnaean Society and a founder member of the Entomological Society of London. Clearly he was a very clubbable and popular man, and his vein of humour found an outlet not only at meetings of these societies, whose members he would often reduce to helpless laughter, but also in the many humorous pieces he would write for *The Entomological Magazine* under such mock-scholarly titles as 'Entomological Sapphics' and 'Colloquia Entomologica'. The kind of humour they displayed went down well with their intended audience, but was of its time and, like much Victorian comedy, now seems ponderously jocose. However, such writings, along with the doggerel often composed to commemorate entomological jaunts and jollities, give an authentic flavour of the hearty comradeship and *joie de vivre* of those cheerful naturalists. Like so many Victorians, Newman had a formidable energy and a tremendous capacity for work, rising as early as three in the morning during summer to gain extra hours for uninterrupted writing or tending to his collections. He was an indefatigable hill-walker and (always a good sign) a keen cricketer, belonging to two clubs and excelling with the bat. Brought up as a Quaker, he was unselfishly public-spirited, for many years sacrificing one day a week to working on the collections of the Entomological Club. Of all the eminent Victorian naturalists, Edward Newman comes across as the one you would most happily spend time with, and his fellow entomologists seem to have felt much the same.

The Victorian Age: Men with Nets

The country the Victorian butterfly men knew was very different from the one we know now. Mechanised agriculture had made little or no impact on the rural landscape, and the sheer abundance of butterfly life was beyond anything we can easily imagine today. I have written about this in the chapter 'My Butterfly Life', and given a few examples of the kind of prodigious – to modern sensibilities, appalling – hauls routinely netted by collectors. But even those with no particular interest, taking a country stroll or enjoying their suburban gardens, could hardly have failed to notice that, on any reasonably sunny day, there were butterflies everywhere: a profusion that was, like so many things, taken for granted until, little by little, it was gone. The great majority of collectors, too, took it for granted, assuming that, however many specimens they netted, there would always be more. For the net-happy butterfly enthusiast, this was a golden age, and Britain was the place to be.

There was much for those energetic Victorians to do, as parts of the country – Scotland, the Lake District, even butterfly-rich Dorset – had only recently been explored by lepidopterists, and the list of British species was still quite blurry at the edges. There were genuinely new species to be discovered, and other supposed native species to be dropped from the list. There was also the complicating factor of fraud by dealers and collectors passing off foreign specimens as having been netted on British soil (as some of them indeed were, but only after they had been clandestinely released). A group of dealers operating in and around Canterbury became known as the 'Kentish buccaneers', so bold and persistent – and often successful – were their attempts to pass off Continental specimens as British, importing them across the Channel as live butterflies, breeding them in their own cages, or releasing them to be caught by the wishful-thinking. They knew there was money to be made by selling these desirable rarities to collectors, and they tried to gain respectability by having their supposed finds officially recognised. Edward Newman, as editor of *The Entomologist*, was obliged to spend much time fending them off, but in the end one of them overreached fatally by reporting that a specimen of the Arran Brown, a northern mountain species, had been taken at Margate – a claim so far-fetched that it finally destroyed the credibility of the Kentish buccaneers.

The activities of these dubious dealers point to another facet of the great Victorian collecting boom. The obsessive acquisitiveness that butterfly collecting, like any other form of collecting, can unleash, the compulsion to acquire some especially elusive rarity, to complete a set, or possess every species. Many collectors fancied having a couple of rarities in their collection, and were happy to pay for apparently genuine specimens without inquiring too closely into their origins. For all the companionable bonhomie of the Victorian collecting world, there was inevitably an element of competitiveness which could result in squabbles and feuds, and sadly, in one case, a lawsuit. The suit was brought by J.F. Stephens (mentioned above as a regular host of entomological 'at homes') against James Rennie, a Scots entomologist. Stephens claimed that Rennie had pirated material from his (Stephens's) *Illustrations of British Entomology* (1828) for use in his own *Conspectus of the Butterflies and Moths of Britain* (1832). The case caused quite a stir in entomological circles and beyond, and, though Stephens lost in court, he was saved from financial ruin by Newman and other sympathisers, who set up a fund to pay his costs. The verdict in the case, Newman remarked, 'reflects anything but credit on the laws of this country'.

Stephens was, even by the standards of the time, an indefatigable cataloguer of the natural world. When he was only sixteen, he embarked on a complete list of all British animals, and four years later he had produced a 'catalogue' of all the British insects known at the time, including 1,367 butterflies and moths. By the time this list had evolved into his *Systematic Catalogue of British Insects*, the number had risen above ten thousand, some two thirds of which Stephens had in his own collection. All this was achieved with virtually no outside assistance, and while working first as a clerk in the Admiralty Office and then at the British Museum. His *magnum opus*, the work of eighteen years, was the twelve-volume *Illustrations of British Entomology* (1828), four volumes of which are devoted to butterflies and moths. Sadly, when he returned from the British Museum to the Admiralty Office, Stephens found that the publicity resulting from the lawsuit had gone down badly with his superiors, and after a while he resigned and returned to the museum, working as an unpaid assistant until his relatively early death, at the age of sixty.

A notably combative character among the Victorian butterfly men was James William Tutt, a schoolmaster and amateur entomologist whose phenomenal energy was such that he found it impossible to be idle, still less to relax. Unlike many of his fellow collectors, he had no private means, and had had to make his own way in the world, rising through the education system to become a school headmaster. His appetite for work, his prickly personality and lack of social graces might be explained by this arduous path through life, and his didactic, sometimes dogmatic style by his having been so long a teacher. His manner was so offensive to some that two of the greatest butterfly men of his day, Richard South and F.W. Frohawk, even resigned their membership of societies simply to avoid his company. 'You know I am often brutal in the way I put things,' Tutt is reported to have said, 'but I cannot help it, and you know I am right.' He died in his early fifties, probably from a stroke, and left no fewer than twenty-four properties, all but one freehold, in his will. It was a remarkable property empire for a not very well-paid schoolmaster.

Richard South could hardly have been more different from Tutt. A popular figure, genial, urbane and a friend to all (with the forgivable exception of Tutt), he was comfortably buffered by private means. Devoting much of his time to collecting and studying butterflies and moths, South was a member of all the main entomological societies, edited *The Entomologist*, and he wrote. Most importantly, he wrote two of the best and most successful field guides ever published, the Wayside and Woodland volumes on *Butterflies of the British Isles* and *Moths of the British Isles* (1907). South and Frohawk were the two giants of the late Victorian and early twentieth-century butterfly world. Frederick William Frohawk was, like South, popular and widely admired. Though he was born into a well-off family, his father's early death left them in straitened circumstances, and the young Frohawk lost nearly all the sight in one eye as a result of typhoid – a circumstance that did nothing to hold him back in his artistic and lepidopteral endeavours. Less interested in collecting than in breeding butterflies, he devoted many years to recording each stage of the life cycle of all the British butterflies – a hugely ambitious project that culminated in the publication of one of the greatest of all butterfly books, his two-volume *Natural History of British Butterflies* (1924). Of this, and of South's field guides, more in the next chapter.

Frohawk was a vigorous outdoor type who thought nothing of walking twenty miles. He invariably wore a Norfolk jacket, the classic country wear of the Edwardian gentleman (in Frohawk's case, unusually, cut square at the neck). In the caricature image that has proved surprisingly long-lived, the butterfly collector is more than likely to be wearing a tweed Norfolk jacket, with a countryman's flat cap, and, of course, a net in his hand, which he will be waving about in a more or less deranged manner. Even in the golden age of butterfly collecting, an aura of eccentricity attached itself to the collector, and continued to stick with him. A much later collector, Vladimir Nabokov, wrote about the range of reactions he had himself experienced to the bizarre spectacle of a man with a butterfly net: from Spanish villagers left 'frozen in the various attitudes my passage had caught them in, as if I were Sodom and they Lot's wife' to suspicious authority figures, Americans howling their derision from passing cars, sleepy dogs suddenly perking up to snarl at his passing, and children pointing him out to their puzzled mothers. As he notes, he was in his forties when he came to America, and 'the older the man, the queerer he looks with a butterfly net in his hand' – which is probably true, especially if, like Nabokov, he is wearing an unbecoming combination of shorts and knee-length socks.

The image of the butterfly collector with his net has generally been one of benign, if unaccountable, eccentricity, perhaps not far removed from lunacy, but not dangerous. On the other hand, there have been some sinister butterfly chasers too, if only in fiction. Jack Stapleton in Arthur Conan Doyle's *The Hound of the Baskervilles* (1902), the cunning criminal plotting to kill off his Baskerville relations, is a keen lepidopterist, forever pursuing butterflies over the treacherous Grimpen Mire. Doyle describes him chasing crazily after his quarry – 'It is surely Cyclopides' (a genus of Skippers to be found in southern Africa, certainly not on Dartmoor) – and later Doyle writes of Stapleton running wildly, 'his absurd net dangling behind him'. Frederick Clegg, the psychopathic loner in John

Fowles's *The Collector* (1963), is an obsessive butterfly collector, an exemplar of what the author, a man of strong opinions, regarded as a 'lethal perversion' (though – or perhaps because – he had once been a keen collector himself). In the mystifyingly popular 1970s television sitcom *Butterflies*, the pathologically buttoned-up husband Ben (played by Geoffrey Palmer) was not only a dentist but, worse, a butterfly collector. Even Enid Blyton, in *Five Go to Billycock Hill* (1957), has a pair of decidedly 'queer' butterfly collectors who share a house and are so wrapped up in their obsession that they fail to notice the heinous crime that is going on under their noses. And then, more recently, came a film named after a blameless butterfly, *The Duke of Burgundy* (2014), which depicts a distinctly creepy sadomasochistic (or, in the jargon, 'sub/dom') relationship between a lepidopterologist and her student. We have come a long way from those amiable gentlemen in their Norfolk jackets.

The title of this chapter is 'Men with Nets', and there is no denying that the Victorian butterfly world was an overwhelmingly masculine realm. But there had always been women with a keen interest in butterflies, and a few of them became active collectors: one of the earliest and most interesting, Eleanor Glanville, has already made an appearance in the chapter *Early Times*, and other butterfly-loving ladies, from the age of the Aurelians, in the following chapter. Generally, as respectability took hold in the nineteenth century, it was considered permissible for a lady to take an interest in entomology, preferably butterflies, but not to go out chasing them around the countryside as freely as the men did. Typically, ladies would breed butterflies and observe their life cycles, and often they would paint and draw them. In Regency England, Laetitia Jermyn, the daughter of an Ipswich printer, was free to botanise (another respectable pastime) and to study butterflies. She was encouraged in the latter pursuit by a Suffolk neighbour, the parson-naturalist William Kirby, whose ground-breaking *Introduction to Entomology* (1815–26) gave him a fair claim to the title 'father of British entomology'. Laetitia, 'the fair Aurelian', as she liked to call herself, produced a delightful book, *The Butterfly Collector's Vade Mecum: or a Synoptical Table of English Butterflies* (1824), dedicated to her mentor, Kirby. The attractive frontispiece, showing a Swallowtail butterfly with caterpillar, chrysalis and food plant, is signed by her, though her name does not appear on the title page. In her preface, she defends butterfly collecting against those who would attack it as 'a trifling and worthless pursuit', and her 'synoptical table' lays out much useful information about butterfly food plants, times of emergence, and likely localities for finding them.

Emma Hutchinson, the wife of a Herefordshire clergyman, came late to lepidoptery, inspired by the delicate beauty of a Swallow-Tailed Moth that her five-year-old son had caught. She very soon became an indefatigable breeder of butterflies and moths, rearing larvae with rare dedication and success. Her speciality was the Comma, a butterfly that at the time was in decline in much of England, but which still abounded in Herefordshire. It was through her work that the Comma was shown to be double-brooded, with the spring-bred generation noticeably paler than the later brood. Fittingly, the pale form still bears the name of Emma Hutchinson: it is known as ***Polygonia c-album hutchinsoni***. She left behind her a collection of over twenty thousand Herefordshire Lepidoptera, which her daughter gave to the Natural History Museum.

A later and very much more colourful figure was Margaret Fountaine, who has been described (by Michael Salmon) as 'one of the strangest lepidopterists Britain has ever produced'. A Norfolk clergyman's daughter, she was brought up conventionally enough, but, having had her heart broken by an Irish singer whom she pursued all the way to Ireland, she left England to travel in search not of love but of butterflies – though along the way she found both. Her pursuit was to lead her through various parts of Europe, Africa, the Middle East, India, the Far East, the Americas and the Antipodes. A dauntless traveller who seemed to fear nothing and to thrive in the most basic living conditions, she probably covered more ground than any lepidopterist has ever done, even in the age of jet flight. Though she sent frequent reports to the entomological journals, it was her other, more private writings that were to secure her posthumous fame. When she died – in 1940, while hunting butterflies in Trinidad – she left her collection of some twenty-two thousand specimens to Norwich Castle Museum, and with it a mysterious black box which, she stipulated, was not to be opened until April 15th,

1978. The day came, the box was duly opened, and it was found to contain twelve thick volumes of her journal, covering six decades and more of her life – and containing an astonishingly frank account of her emotional life, notably her relationship with her Syrian dragoman (guide, interpreter and, in Margaret Fountaine's case, a great deal more). Khalil Neimy had fallen in love with her on their first meeting, despite being fifteen years her junior, and he became her 'dear companion, the constant and untiring friend' on her travels for many years. They exchanged rings and planned to marry and settle in America, but Khalil sadly died on a return visit to Syria in 1929. Wisely, in view of the attitudes that prevailed at the time, they kept their feelings concealed from the world, and Margaret Fountaine equally wisely chose to keep her intimate diaries under embargo until the hundredth anniversary of the date on which she began writing them. When their contents became known, they caused a stir far beyond the world of lepidoptery, and their contents were adapted into two successful books by W.F. Cater: *Love among the Butterflies* (1980) and *Butterflies and Late Loves* (1988). Margaret Fountaine became, along with Eleanor Glanville, one of the two romantic heroines of butterfly collecting – two more than there are male romantic heroes.

When she was collecting, Margaret Fountaine favoured practical but somewhat eccentric costumes of her own devising, with loose, ankle-length dresses, cotton gloves with the tips of the thumb and index finger cut off (perhaps to enable her to dispatch her catches by crushing the thorax), a compass on a chain slung from a buttonhole, and a cork sun helmet. This, of course, is not the image of the butterfly collector that has come down to us from the Victorian age. The stereotyped image is entirely masculine – a man in a Norfolk jacket and flat cap, brandishing a net. And there is invariably one more detail: across his chest is the shoulder strap of a capacious bag, in which would be, as likely as not, a field guide, specimen tins and other bits of butterfly paraphernalia. These essential adjuncts of the collecting life, products of a once booming industry, will be one of the subjects of the next chapter.

Pins, Poisons and Field Guides

As the love of butterflies – finding expression, increasingly, in the urge to collect them – spread through all levels of society and into every part of the country, a small industry grew up to serve the needs of all those butterfly fanciers, from the casual net-waver to the serious lepidopterist. Now that the pursuit of butterflies was no longer limited to a London-based coterie of relatively prosperous, aesthetically inclined enthusiasts, more practical, portable and inexpensive equipment was required. When the Aurelians went collecting, they could store all they caught and all they needed – even their nets – in the deep, capacious pockets of their frock coats, and could pin specimens to the outside and inside of their tall hats. The Victorian collector, relatively underdressed by eighteenth-century standards, had much less storage capacity about his person, and needed two things: a bag, and more compact and convenient equipment.

First, the net. In the course of the Victorian period, the Aurelians' cumbersome, two-handled 'clap-net' or 'bat-fowler', which looked like some kind of trawling net or scoop, was gradually replaced by simpler forms, with a more or less round net on the end of a single handle (some models could be attached, handily, to a walking stick). These simple balloon or kite nets, as well as being easier to use, could be dismantled and stowed in the collector's bag. So much more practical was this style that even the most conservative or antiquarian-minded of collectors had abandoned the clap-net by the end of the nineteenth century, and today it seems that not a single original clap-net survives, even in a museum. Being endlessly inventive, the Victorians came up with plenty of variations on the balloon or kite net, including one that could be passed off as an umbrella (until it rained, anyway). The most striking variants were the 'high nets' designed to catch those species, notably the Purple Emperor, that haunt the treetops and seldom come down to the ground. An early model, devised by the Reverend F.O. Morris (a strong opponent of Darwin and author of several popular handbooks), rose to a height of fifty feet, with shrouds and stays reminiscent of a ship's mast. A more usual height was around fifteen feet, but some collectors, including the hospitable Stainton (last sighted in the previous chapter), favoured a length of thirty or forty feet, which must have been extremely cumbersome both to use and to carry. Since the aim of the Victorian butterfly lover was to collect specimens, he needed, as well as collecting boxes for temporary storage, the means to kill, pin and set the unfortunate victims of his passion. Crushed laurel leaves, which release prussic acid, became a popular means of killing specimens, though cyanide was still being recommended in butterfly books well into the twentieth century, along with chloroform and liquid ammonia. The butterfly had then to be pinned to a setting board and its wings carefully moved into the desired, fully open position until rigor mortis completed the job. This was a delicate process, requiring considerable skill, as I remember, with shame, from my misspent boyhood. Then, for the serious collector (which I never was), a suitable cabinet or cabinets, preferably of mahogany, would be required to house those drawers full of dead butterflies. One firm came to dominate the business of supplying all those collectors with everything they needed – Watkins & Doncaster, who in 1879 set up shop on the Strand in London. Arthur Doncaster, co-founder of the company, was profoundly deaf and speech impaired, and communicated with customers by means of a slate hung around his neck. When a customer asked for something, he would write his answer on the slate and pass it to them to read and hand back. The business was not confined to entomology: when glass cases full of stuffed birds became all the rage in the late Victorian period, Watkins & Doncaster employed five taxidermists to stuff and mount suitably decorative birds. The company left the Strand in 1956, but continued to operate from various locations, including a shop in the Lanes in Brighton. It is now based in Herefordshire and continues to thrive, though the nature of its business and its merchandise has of course changed greatly since Victorian times.

Another essential piece of kit was the field guide, preferably small enough to fit into the pocket. As the beautifully illustrated deluxe edition or, as we might now call it, coffee-table book, was the characteristic product of the Aurelian period, so the practical field guide, primarily an aid to identification, was that of the Victorian-Edwardian age. It took a while to evolve. Works of the early nineteenth century such as A.H. Haworth's comprehensive *Lepidoptera Britannica* (1803–28) and John Curtis's beautifully illustrated, sixteen-volume *British Entomology* (1824–39) were as cumbersome as they were compendious, and aimed squarely at the gentleman scholar with ample library space. However, more condensed (and cheaper) works, such as William Kirby (Laetitia Jermyn's mentor) and William Spence's *Introduction to Entomology* (1815–26), aimed at the beginner, and Stainton's *Manual of British Butterflies and Moths* (1857), published in monthly parts, were paving the way for the compact field guides to come. Another sign of growing interest in and affection for butterflies, and insects in general, was a curious sub-genre of children's insect books, aimed less at practical use than at moral instruction. Some of these were published by the Religious Tracts Society and presented the wonders of Nature as most definitely the handiwork of God, and the metamorphosis of insects as a potent image of the resurrection into eternal life. One of the most popular, though, was the purely whimsical narrative poem *The Butterfly's Ball, and the Grasshopper's Feast* by William Roscoe, a work that enjoyed a surprising new vogue in the 1970s when William Plomer and the fashionable illustrator Alan Aldridge adapted it to contemporary taste.

As printing became cheaper and faster, thanks to steam-driven presses and the lifting of the tax on paper, and as lithography began to take over from wood and copper engraving, allowing the production of high-quality colour plates at relatively low cost, books and periodicals were coming within the reach of more and more buyers.

At the same time, elementary education was spreading, illiteracy was falling, and a whole new public for books was coming into being (think of Mrs Gaskell's hand-loom weavers). Suddenly natural history books aimed at this public, though often of dubious quality, were selling in huge numbers. One of these, the long forgotten *Common Objects of the Country* by the Reverend J.G. Wood, published as a shilling handbook in 1858, sold an astonishing one hundred thousand copies in its first week of publication. Even such famous Victorian bestsellers as Samuel Smiles's *Self-Help* or *Mrs Beeton's Book of Household Management* came nowhere near those sort of figures. There was scope, too, for works of higher quality produced at an affordable price, and the great Edward Newman produced several successful works along these lines, including his much-praised *Illustrated Natural History of British Butterflies*, published in a single handsome volume in 1871. The Reverend F.O. Morris had a success with his *History of British Butterflies* (1853), a handily sized octavo volume with coloured plates which was very popular in its day, though its ponderously wordy and whimsical style makes it all but unreadable today. Henry Guard Knaggs, whom we have met before, produced a breezily written, practically useful volume, *The Lepidopterist's Guide* (1869), which remained in print for some forty years. But the best and most successful field guide, the one that was most practically useful and that effectively set the template for all future field guides, was *Richard South's The Butterflies of the British Isles: A Pocket Guide with Descriptive Text*. This was first published in Frederick Warne's Wayside and Woodland Library in 1906, and proved so popular and easy to use that it was still in print in the 1960s. Its companion work, the two-volume *Moths of the British Isles*, was even longer lived, being reprinted as late as the 1980s; both works were regularly revised and updated. South's book was among the first to use colour photographs of all the butterflies, as well as high-quality drawings of the earlier life stages. Pocket-sized and packed with information about each butterfly's appearance, habits, life cycle and distribution, it is also very readable, even today. The charming Preface, though a little whimsical, breathes the spirit of enchantment:

'Almost everyone admires the wild flowers that Nature produces so lavishly, and in such charming variety of form and colour; but, in addition to their own proper florescence, the plants of woodland, meadow, moor or down have other "blossoms" that arise from them, though they are not of them. These are the beautiful winged creatures called butterflies, which as crawling caterpillars obtain their nourishment from plant leafage, and in the perfect state help the bees to rifle the flowers of their sweets, and at the same time assist in the work of fertilisation.

It is the story of these aerial flowers that we wish to tell, and hope that in the telling we may win from the reader a loving interest in some of the most attractively interesting of Nature's children.

South goes on to acknowledge that (even in his day) there are some who would sooner simply watch and learn about these creatures, rather than kill and collect them. 'It is believed,' writes South, 'that this little volume will be found useful to both sections of naturalists alike.' It was – and is – indeed, even if it is no longer the book that the butterfly lover would take with him into the field.

The towering achievement of this period, though its publication was delayed until 1924, was F.W. Frohawk's *Natural History of British Butterflies.* Published in two volumes at a prohibitive price of six guineas (more than six hundred pounds in today's money), this work, with its exquisite colour illustrations, was in a sense a throwback to the sumptuous volumes produced by the Aurelians. However, unlike them, it was a comprehensive and authoritative reference work, and those beautiful illustrations were the product of years of close study and minute observation of every stage of life of all the British butterflies, every one of them reared by Frohawk himself. 'Twenty-four years of unremitting research,' wrote Frohawk feelingly, 'were absorbed in the course of the author's observations.' It is an astonishing achievement, and the colour plates are as beautifully drawn and designed as they are informative. Happily, Frohawk's ***magnum opus*** was repackaged in affordable single-volume form as *The Complete Book of British Butterflies* (1934), which became a standard work.

Many practical field guides were to follow where Richard South had led, culminating in more recent times in the work of lepidopterist Jeremy Thomas and illustrator Richard Lewington. Their collaborative work ***The Butterflies of Britain and Ireland*** (1991), illustrated throughout with Lewington's superb artwork, is a volume as beautiful as it is authoritative, and the pocket-sized guides written by Thomas (which use mostly photographic illustrations) and Lewington (all watercolours) are the most useful and usable of all current field guides. It is rather wonderful that, in an age when photography is supposed to have superseded painted illustration, paintings of butterflies (especially if they are of such a high standard as Lewington's) are still the best way to represent the butterfly, its life stages, habitat and food plants – not least because all these elements can be brought together in one plate, which of course could not be done by photography. The artist also has more choice than the photographer, even in the age of digital manipulation, over how he presents his subject, in terms of pose, viewpoint, lighting. The knowledgeable illustrator can design his single image to emphasise exactly what needs to be shown for identification purposes, and to convey the overall 'feel' of its subject. Even the cleverest photography captures, by definition, a snapshot; a good painted illustration can capture an essence. The very stylisation of the handmade image allows it to be, in a sense, more real than the 'real thing' caught by the camera lens.

While all those amateur butterfly lovers were in the field, with their nets and collecting tins and pocket guides, the science of lepidoptery was advancing on another front, as the Victorian age saw the birth of scientific specialism, of laboratory-based science and the development of procedures, specialised equipment and professional jargon that were largely inaccessible to the lay person. The scientists had, in a sense, retreated from the field into the laboratory, separating themselves from the endeavours of amateurs and developing their own arid, precise and affectless language and attitude. Scientific materialism was seen by the new scientists as a sufficient account of the whole nature of reality, and the whole reality of nature. This appalled, among others, the great Victorian critic and sage John Ruskin, who reacted in horror to an essentially reductionist science that, rather than enlarging the human world, sought only to diminish it. Science seemed to him to have become inimical to art, and ultimately to humanity, undermining man's ability to benefit from an emotional and aesthetic relationship with nature. Field naturalists, however, were generally taking things more calmly: Darwin's revolutionary idea of natural selection struck many entomologists as unexceptionable, perhaps because Darwin, an obsessive beetle collector, was a 'bug man' like them. Others, like the Rev. F.O. Morris, regarded his ideas as obviously absurd, and launched broadside after broadside against them. 'The entomologists,' Darwin grumbled in 1863, 'are enough to keep the subject back for half a century' – and indeed it was not until the 1880s that the Entomological Society of London accepted Darwinian natural selection. Meanwhile, for most nature lovers, the evangelical urge continued stronger than any science, and popular writers happily talked of 'the God-written Poetry of Nature'.

Even in this time of rapid scientific advance, enlightenment could not be wholly disentangled from enchantment. If pure scientific curiosity drove the specialists' activities, the allure of butterflies for most people was more emotionally grounded and more strongly aesthetic. This aesthetic delight spilled over into the decorative arts, with butterfly brooches and pins becoming extremely popular with Victorian ladies, and die-printed cut-outs of butterflies (few of them at all accurately depicted) being widely used in scrap-books and decorative collages. One of these cut-outs turns up in the unlikely context of a publicity photograph of the American poet Walt Whitman, posing with what he claimed was a living butterfly perched on his fingertip. 'I've always had the knack of attracting birds and butterflies and other wild critters,' he told the historian William Roscoe Thayer, who later commented drily: 'How it happened that a butterfly should have been waiting in the studio on the chance that Walt might drop in to be photographed, or why Walt should be clad in a thick cardigan jacket on any day when butterflies would have been disporting themselves in the fields, I have never been able to explain.' In fact, as can be clearly seen, this butterfly is no living 'critter': it is a cardboard cut-out, attached to the poet's finger by a ring. This curious relic is now preserved in the archives of the Library of Congress.

In the Victorian home, the decorative possibilities of butterflies were exploited to the hilt: the more spectacular species were displayed in cases and frames and under glass domes like so many tropical birds, all for purely decorative effect. There was even a craze for clipping off butterflies' wings and mounting them in albums as if they were stamps, painting in the bodies between pairs of wings: butterfly mania as a grisly variant of philately. The demand for butterfly specimens for use in interior décor combined with the inroads made by the collectors created a situation in which it became apparent to some, at least, that this plundering could not go on indefinitely, even in an age of abundance. With many collectors amassing dozens of specimens of a single species, and dealers anxious to supply the market, sometimes taking hundreds of specimens from single isolated colonies, the

Entomological Society of London decided in 1896 that it was time to pay serious attention to the problem of over-collecting. James Tutt (whom we have met before) recorded a 'remarkable discussion' that took place at a meeting of the society in May of that year. 'There can be no doubt,' he wrote, 'that, in the opinion of many lepidopterists, the man who simply collects is rapidly becoming a public nuisance, and it is pretty well understood that his exterminating processes act distinctly as a check to the scientific aspirations of entomologists.' This is a plea for conservation in the cause of scientific study rather than any larger considerations, but it marks the beginning of a movement away from unrestrained collecting and towards conserving our butterflies. In the same year, the Entomological Society set up a committee 'to consider the protection of British insects in danger of extinction', and the following year this committee recommended the formation of an association specifically designed to discourage the over-collection of butterflies and moths. Though it might have been little noticed at the time, this was a significant early indication of a change of mood in the collecting world, and these initiatives in the 1890s paved the way for the conservation efforts of the following centuries, and the philosophy of conservation that was to become dominant across the board.

The Twentieth Century: Conservation, Citizen Science, New Wonders

As the nineteenth century drew to a close, it began to seem as if the butterfly world – along with the whole field of 'natural history' – was in danger of dividing into two tribes: the amateur naturalists out in the field collecting and enjoying, and the laboratory-based professionals, hunched over their microscopes or pondering vexed questions of taxonomy. However, in the following century the two tribes were to draw closer together as the focus of scientific interest shifted in the direction of conservation, a project that demanded feet on the ground, and as many observant eyes as could be mustered. This was part of a wider movement, the result of a growing perception of serious threats to species and to habitat. One arm of it was the amenity movement that sought to conserve countryside and open spaces – the original function of the National Trust (founded 1895) – and to create nature reserves protected from harmful human activity. Another arm was the powerful movement to protect birds – the first conservation movement to achieve major success in Britain, with the founding of the Society for the Protection of Birds in 1889 and its ultimately successful bid to ban the import of exotic plumage to decorate ladies' hats ('murderous millinery'). The organisation Butterfly Conservation was not formed until 1968, but, as we have seen, many butterfly lovers were already aware of the need to protect species and habitat even before the end of the nineteenth century.

By the middle of the twentieth century, a new kind of amateur naturalist was entering the field: more scientifically aware, less concerned with hunting and collecting, and equipped with more advanced, affordable and easy-to-use cameras, binoculars and other optical devices, making the netting of specimens less necessary as well as less desirable. Butterflies were now seen less as quarry or collectibles and more as objects of scientific interest in themselves, and, increasingly, as creatures facing existential threats from which they must be protected. The rise of this new, scientifically literate breed of naturalists was recognised by the publishers Collins, who in 1946 launched the New Naturalist series, the first volume of which, fittingly, was *Butterflies* by E.B. Ford. This begins with a very readable history of butterfly collecting in Britain, but sadly the rest of the volume is dry and dauntingly technical, devoted to the study of structure, classification, butterfly genetics and evolution, with copious statistics and diagrams.

That it is so dry is a surprise, as E.B. 'Henry' Ford was a notably flamboyant and eccentric man. A 'confirmed bachelor' and living caricature of the Oxford don, Ford was one of the most notorious University 'characters' of his time. He flaunted the worst kind of misogyny, disparaging 'female women' at every opportunity (though he made an exception for his fellow entomologist Miriam Rothschild). If only women – no men – turned up for a lecture of his, he would look around, declare 'Nobody present' and walk out. He affected a high patrician tone, reacting to his first sight of a fish-and-chip van with the declaration, 'I would not have believed it possible that there could be so much wickedness in the world.' He also strongly disapproved of newspapers and the radio, invariably used the terms 'photographic camera' and 'cinematic projector', and called any piece of apparatus required for the laboratory 'the engine'. But Ford also had a taste for *Carry On*-style innuendo, raising his hat to a friend's nanny whenever he called, and inquiring, 'How is your pussy?' He ate moths (ostensibly as part of his research), talked to himself, affected strange walks, and would speak of 'my friend the Pope'. And every summer, he would lead his young researchers on a pub crawl, tasting all the beers of Oxford, cleansing his palate with a gin after each pint, and remaining to all appearances absolutely sober. So *Butterflies*, it seems, was written by the Kenneth Williams of zoology – who would have guessed? If only some hint of his character had got into his prose, it would have been a very different book.

But, to return to the subject, another sign of the coming together of laboratory science and natural history in the field was the development of various forms of co-operative research dependent on amateur participation: wildlife censuses, network research (drawing together the work of different groups) and 'citizen science', a prime example of which is Butterfly Conservation's annual Big Butterfly Count, in which cause many thousands of members of the public (more than 85,000 in 2024) count the butterflies they see over a given period at the height of the butterfly season. In addition to this eye-catching initiative, much work is done every season by butterfly-loving amateurs walking transects (defined paths from A to B) and noting every butterfly and day-flying moth they see. All of this, as well as being of local interest, feeds into larger scientific endeavours and creates a far broader picture than could ever be made by isolated local initiatives alone.

The rise of the ecology movement and growing environmental awareness from the 1960s onward have undoubtedly coloured our appreciation of the natural world and the threats it faces, as has increasing anxiety about the climate, to the point where it often seems that any news item relating to butterflies will be a tale of loss and decline. There is truth, alas, in these dismal stories, though it is seldom the whole truth, which tends to be rather more complex once it is examined in any depth – but the tone is set: if it's butterfly news, start tolling the bell; it's going to be bad. (To be fair, there was an exception to this rule in 2023, when the results of that year's Big Butterfly Count were, surprisingly, headlined 'Butterfly numbers increased this summer', a rare good news story that was taken up in other media.) This, taken in combination with the methodical, number-crunching approach that now prevails (necessary though that is), tends to eclipse another important aspect of the study of butterflies – the enchantment that these creatures can weave, the sheer pleasure they give to the watcher, and the wonder they inspire. And that wonder is also to be found, increasingly, in a less likely context – in the science itself...

As science made dramatic strides in the nineteenth century, there was some anxiety, particularly among Romantic poets and thinkers, that scientific investigations would disenchant our world, 'unweave the rainbow' (Keats's phrase), and undermine the sense of wonder and mystery that we instinctively feel in the face of Nature. There were good reasons for this anxiety, as an increasingly mechanistic view of Nature's workings seemed to be well on its way to explaining away everything. However, things were never that simple: Darwin himself, whose evolutionary theory did so much to shake the old sense of a meaningful, God-infused world, was driven from the start by a lively sense of the beauty and wonder of the natural world. In an early sketch of his overarching theory, reused later in *The Origin of Species* (1859), he wrote rhapsodically that 'there is a simple grandeur in the view of life, with its powers of growth, assimilation and reproduction, being originally breathed into matter under one or a few forms, and that whilst this our planet has gone circling on according to fixed laws, and land and water, in a cycle of change, have gone on replacing each other, that from so simple an origin, through the process of gradual selection of infinitesimal changes, endless forms most beautiful and wonderful have been evolved.' 'Simple grandeur', 'breathed into matter', 'most beautiful and wonderful' – this is hardly the language of scientific materialism. And today, looking back from where science has got us to now – after Einstein's transforming insights, the mind-boggling vistas opened up by quantum physics, the dramatic findings of genetics, and the new worlds revealed by electron microscopy – we can surely say that the realm of wonder, even of beauty, has not been shrunk by science but vastly widened and given more substance. The more we look at Nature with the instruments and insights of science, the more wonders and mysteries open up; and the more we learn, the more we discover there is to learn. The scientific mind seems to live happily enough with this, working on the assumption that all mysteries will in the end be solved, that all can and will be explained, that there is an end point. But is there any good scientific reason to believe that? It seems to be more a matter of faith; it might well be that what lies before us is an endless vista of mysteries within mysteries, wonders upon wonders.

We used to think we knew about butterflies. This was largely because we had investigated them too little to realise the limits of our knowledge. The science had tended to focus more on taxonomy, on matters of naming and classification, than on the basics of butterfly behaviour, and on how a species – indeed an individual – lives in its environment. Even such a basic question as how a butterfly, with its small body and cumbersome expanse of wings, manages to take off and fly away so fast and effectively (at first sight of a camera, in my experience) was not answered until very recently. Analysis of slow-motion photography, publicised in the *Journal of the Royal Society* and the

Smithsonian Magazine, has revealed that, on take-off, a butterfly 'claps' its wings together, flexing them and creating an air pocket that gives it extra thrust, a touch of 'jet propulsion'. And it's gone. Then there is the question of how such a relatively heavy butterfly as the Red Admiral can fly so fast, manoeuvre with such agility, and make long-distance migratory flights. Research using a wind tunnel and slow-motion photography revealed an extraordinary range of wing movements that create vortices and turbulent airflows over the butterfly's wings, enabling uplift, propulsion and swift, agile movement far beyond anything that might have been expected.

The extent of our ignorance about butterflies was revealed dramatically in the course of a campaign to save the Large Blue, a species that was always a prized rarity, but which, by the 1970s, was on the very brink of extinction in England, reduced to a few small, discrete colonies in Devon and Cornwall. It could not, in the event, be saved from its sad fate (it became officially extinct in the UK in 1979), but this still became perhaps the most famous success story in British butterfly conservation. It was the great lepidopterist Jeremy Thomas who first worked out (just too late) what had caused the extinction. It was not human intervention but rather the lack of it: the Large Blue needed, first and foremost, short-cropped turf, with the grass at just the kind of height (under three centimetres) produced by carefully managed grazing. When the last known colony of Large Blues had been fenced off from grazing livestock and intruding humans to 'protect' the butterflies, this well-meaning measure had in fact condemned them to extinction.

It was known that the Large Blue needed wild thyme, its larval food plant, and that the larvae had a relationship with ants, but what was not fully known was the mind-boggling complexity of that finely balanced relationship. When a Large Blue larva hatches, it begins its life feeding, like a good vegetarian, on the flowerheads of the wild thyme, but then it throws itself onto the ground and starts secreting a fluid that seems irresistibly attractive to red ants of the species *Myrmica sabuleti*, which cluster round in a feeding frenzy. At some point the caterpillar will raise itself into an 'S' shape and, at this signal, the ants will carry it into their nest. Once there, the larva ensures its tenure by emitting sounds that, to its subject ants, suggest it is a queen ant, and can therefore do exactly what it wants and still be waited on diligently. Turning carnivorous, the larva immediately sets about devouring quantities of ant eggs and grubs with impunity, fattening itself up for months before pupating for three weeks, then being escorted from the nest by its entourage, emitting regal sounds even as the butterfly breaks free from the chrysalis. This extraordinarily elaborate deception carries a high risk of failure: sometimes too many caterpillars will end up in the same nest, so many of them will starve for lack of grubs; at other times a Large Blue larva will pretend to be queen in a colony which already has a full complement of queens, so the undeceived workers will turn on it and kill it. A further, barely credible complication is that Large Blue larvae can sometimes find themselves hosting a parasite of their own – a rare wasp that enters the ants' nest, sprays around a chemical that causes the ants to start fighting each other, and, in the ensuing mayhem, injects its eggs into the Large Blue larvae. Recovering from this chemical attack, the ants resume feeding the larvae – but what they are feeding now is the wasp grubs growing inside the caterpillars and taking over the entire body cavity. In due course new wasps emerge from what were once Large Blue larvae and, spraying more of the disabling chemical around, head for the exit. The parasite has itself been parasitised. As the mathematician Augustus de Morgan put it (paraphrasing Jonathan Swift), 'Great fleas have little fleas upon their backs to bite 'em,/And little fleas still lesser fleas, and so ***ad infinitum***.'

All this leaves the mind reeling. How could such a complex web of interrelationships have evolved? It is easy enough to see how it could have broken down – but how did it ever become a functioning system in the first place? The Large Blue, it turns out, is not alone in having a special relationship with ants: the larvae of most British blues are known to be attractive to ants, because of secretions from their bodies and, in some cases, sounds they make. However, none, as far as we know, have developed the relationship to such a level of sophistication as the Large Blue. As well as Blues, some of those elusive fascinators, the Hairstreaks, have relationships with ants: the Brown

Hairstreak's pupae are tended by ants, encouraged by the chrysalids' 'chirruping' sounds; the Purple Hairstreak, which 'sings' at both the larval and pupal stage, often pupates in ants' nests; and Green Hairstreak pupae, which make sounds that are even audible to the human ear, are commonly attended by ants. (These sounds are made by 'stridulation', rubbing together the edges of two abdominal segments, one toothed and the other grooved.) Who knows what other intricate relationships might be involved in the lives of other butterflies, many of which – especially the smaller, less conspicuous ones – have not been intensively studied?

No British species, indeed, has been studied in such depth as the Large Blue – and most of that research effort might never have happened if it were not for the imminent threat of extinction, and the timely appearance on the scene of Jeremy Thomas. Why was this the most famous success story in British butterfly conservation, despite the extinction of the species? It was because Thomas's painstaking research revealed such an extraordinary story, and correctly identified what had caused the extinction and what the precise needs of this butterfly were. Once that was known, it was clear that a suitable habitat could be created and, with luck, the Large Blue could fly again in England. With the English strain extinct, Scandinavian specimens, indistinguishable from their extinct cousins, were reintroduced in a few carefully restored sites. They bred, they thrived, and Large Blues now live out their extraordinary life cycle at thirty or more sites in southwest England, notably in the Polden Hills in Somerset, where, in a good year, they fly in their thousands.

In a sense, the great contribution of the Large Blue conservation effort was to expose our ignorance, particularly of the precise habitat needs of species, and the complexities of their relationships with their immediate environment and the living creatures in it. This knowledge has fed into countless later conservation efforts, many of which have been gratifyingly successful, and research into species not in need of any conservation effort has also led to mind-boggling discoveries: consider the Large White, that extremely common bane of brassica-growers, and its relationship with a tiny parasitic wasp. When the male Large White mates with a female, he sprays her with an anti-aphrodisiac chemical to put off other males, but this chemical is highly attractive to a particular species of minute parasitic wasp (so small that twenty adults can fit inside a single Large White egg). This wasp perches behind the female's eye and stays there until she starts laying her eggs, at which point the wasp hops off and sets about injecting her own eggs into those of the Large White, where they feed up and grow into adults. And, as if this wasn't enough, there are at least two other species of parasitic wasps that lay their eggs directly into Large White caterpillars. The resulting grubs gradually eat up the caterpillar from inside, carefully avoiding the vital organs until they are ready to pupate, at which point they kill the hapless caterpillar, emerge and line up in neat rows of cocoons along either side of the caterpillar's corpse. Discoveries like these beg the question – what else do we not know? What wonders (and horrors) might future research programmes uncover?

The greatest wonder of all is the one that has fascinated us humans ever since we first became aware of it – metamorphosis. As with so much else, the more science finds out about it, the more wondrous it seems. For a long time, at least at the level of folk knowledge, people struggled to make the connection between the different stages of a butterfly's life cycle – and no wonder: it is a dazzling feat of natural magic. How could those grubs, those voracious worms with legs, turn themselves into dry, dead-seeming cases, and how could those mysterious casings then split to reveal the living butterfly in all its glory? Naturally we gave human meaning to this process of metamorphosis, seeing in it an intimation of our own resurrection, and, equally naturally, science also got to work, explaining in naturalistic terms what metamorphosis entails and how it happens. Once again, the insights of science make the process seem no less magical and wonderful – especially the stage between the caterpillar pupating and the butterfly emerging. What happens in this metamorphosis, (this 'long enchantment' in the phrase of the American poet Janet Lewis), is that almost every part of the caterpillar dissolves into a kind of goo, a 'butterfly soup', from the constituent parts of which, amazingly, a fully formed butterfly, with its intricately patterned wings, its velvety body, its multi-faceted eyes, its proboscis and delicate antennae, is somehow assembled. This potential butterfly was present all along in the caterpillar, as can be seen by the six front legs, which are more like butterfly legs than caterpillar legs, and the form of the butterfly is hinted at in the chrysalis, particularly in the half-formed 'wing' shapes in its casing. However, the creation of the finished butterfly out of undifferentiated goo still seems little short of miraculous. Furthermore, that goo apparently retains memories from the life of the caterpillar: in experiments, moth larvae trained to avoid a particular chemical have shown the same aversion after they emerge as moths. It could be, also, that larval memory is what tells female butterflies where to lay their eggs – on the plants whose leaves they munched as caterpillars. Whether that is the case or not, it is clear that some form of memory is retained throughout the almost total dissolution that occurs in the chrysalis. Either this is because enough of the nervous system survives to hold some residual memory, or we need to rethink our ideas of what the nervous system is, and what memory is, and how they work.

Butterfly colour is another area in which scientific investigation has uncovered new wonders. We now know, for example, that a surprising range of colours are produced by a kind of optical illusion, created not by pigmentation but by light falling on the scales that cover a butterfly's wings (*Lepidoptera* means 'scale-wings'). Some of these scales are modified in such a way as to scatter light from all wavelengths other than that of one single colour. The most spectacular example is that tropical marvel the Blue Morpho, beloved of all visitors to butterfly houses. Its brilliant, ever changing metallic blue is the product of light being reflected off microscopic structures, reminiscent of Christmas trees or ferns in shape, in the butterfly's wing scales, dispersing and cancelling out all colours but that glorious blue. A similar mechanism creates the radiant emerald green underwing of the Green Hairstreak, a butterfly at the opposite end of the size spectrum from the mighty Morpho. Scientists have lately been taking a particular interest in the kind of scale structure that can produce such a metallic green effect, and have discovered that it is a demonstration in nature of a structure that originated as a purely theoretical entity – the gyroid. This is defined as 'an infinitely connected triply periodic minimal surface' – which is not tremendously helpful to those of us unfamiliar with topology. I can just about understand that the gyroid is a kind of three-dimensional honeycomb structure, and is unique in having triple junctions and no lines of reflectional symmetry, but beyond that my brain refuses to go. However, it is good to know that the Green Hairstreak, one of my favourite butterflies, is unwittingly playing its part in cutting-edge science – science that could, it seems, have applications in computer electronics and anti-forgery logos. In evolutionary terms, this kind of iridescent colouring, changing with the light, presumably has some effect in deterring predators, confusing and dazzling them. The mechanism seems, like so much else in butterfly biology, quite fantastically over-engineered.

Mimicry and camouflage, in butterflies and moths and their larvae, were phenomena that particularly fascinated the young Vladimir Nabokov, as he recalls in his memoir *Speak, Memory* (1951). He identified in lepidopteral mimicry 'an artistic perfection usually associated with man-wrought things'. In some cases, its ultra-refinement seemed to him to defy explanation in terms of Darwinian natural selection: 'nor could one appeal to the "struggle for life" when a protective device was carried to a point of mimetic subtlety, exuberance and luxury far in excess of a predator's power of appreciation'. There are plentiful spectacular examples of such excess in moths and their larvae, and many British butterflies also display such exuberance, from the over-engineered verisimilitude of the Comma's ragged wings – a dead leaf when folded, with a little silver mark exactly mimicking a tiny tear or wormhole letting light through – to the four large, glinting 'eyes' of the Peacock, like the eyes of no earthly creature, almost as unaccountably *de trop* as the tail of its bird namesake. The 'eyes' sported by many less spectacular butterflies have a clear enough apparent purpose (to surprise and/or divert a predator's attention from the more vulnerable body) but the details of them, the perfectly placed 'glint', the delicately outlined rim, the sheer profusion of them in some species, seem to go beyond mere utility. Nabokov was not an anti-Darwinian, but he concluded that such prodigality must be the result not of random point mutations (changes within the gene in which just one base pair in a DNA sequence is altered), but of a relatively sudden large-scale mutation that, as it were, overshot the mark, taking things a great deal further than they needed to go. Those 'eyes' need not so closely mimic the play of light on an open eye, as many butterflies get by with much more basic eyespots, or none. Their perceived resemblance might be no more than an unlikely (and to us, delightful) coincidence, something that Nabokov likens to a happy typographical slip that transforms the meaning of a sentence: 'the chance that mimics choice, the flaw that becomes a flower'.

It is hard not to get the impression of something like playfulness, something like artistry, in the patterning of a butterfly's wings. This is, of course, only a human's-eye view (what else could it be?): we cannot really know how a butterfly appears to its predators, to members of its own species, or to whatever other creatures take notice of it. We can be sure that, seeing the Peacock's wings, they will not be put in mind of the bird's tail, but what will they actually be seeing? More to the point, how does the butterfly, going about its butterfly business, perceive the world? Its eyes, we know, are clusters of numerous miniature eyelets, so its visual idea of the world is an ever-changing compound of multiple images, from which it somehow makes a single sense. We know too that butterfly vision, like that of most insects, extends into the ultraviolet spectrum (and a little way into the infrared), so butterflies will not see colours in the same way as we do. Butterfly senses are much more diffused over their bodies than ours are. Even sight is not necessarily limited to the eyes: Asian Swallowtails, and probably other species, have light receptors in their reproductive organs, of all places.

Such a surprising fact calls for a small digression on butterfly genitalia. These are a subject in themselves – one to which Nabokov devoted a great deal of his career in lepidopterology. He spent many, many hours squinting down a microscope observing the astonishing range and complexity of butterflies' reproductive apparatus. This, in the days before DNA technology, was an important way of distinguishing one species from another, as the male organs of two apparently similar species might be completely different, designed specifically to dock with the female organs of two distinct species. This docking is a delicate, complicated process (hence, presumably, the Asian Swallowtail's curiously placed light receptors) and the apparatus involved is correspondingly complex and richly various. As he stared down the microscope and drew what he saw, Nabokov became fascinated by the fantastic shapes and chance resemblances he observed in butterflies' genitalia, from the raised fists of two boxers squaring up, to men in Ku Klux Klan hoods, a chicken leg, a knife, veiled dancers … Once again, the whole thing seems over-engineered to a quite mystifying extent.

To return to butterfly senses, taste is perceived through sensors in the feet and elsewhere, which enable butterflies to taste the sugar in nectar, or to taste leaves and judge whether they would make suitable larval food. A butterfly's 'ears' are located on its underwings and point backwards, allowing it to pick up the low-frequency sounds of an approaching bird or other predator. Sensory hairs on the butterfly's body and antennae enable it to feel its environment by touch, conveying, among other things, information about the air around it while it is flying. Smell is extremely important to butterflies, especially for finding food and locating potential mates. Accordingly, smell receptors are scattered across much of a butterfly's body, notably the antennae, palps (on either side of the proboscis) and legs. There is much yet to be discovered about butterflies' senses, but we know enough to understand that they inhabit a sensory world quite unlike our own, a world of sights we cannot see, smells we cannot smell, sounds we cannot hear, sensations we cannot feel, chemicals we cannot perceive. And when we see a butterfly in flight, we cannot know what chemical trail it is pursuing, what dense sensory world it is passing through.

The Chinese Taoist philosopher Chuang Tzu one night dreamed that he was a butterfly. As he flew from flower to flower and felt the breezes wafting him to and fro, he found the experience entirely real: he was a butterfly, living as a butterfly lives. When he awoke and returned to himself, he realised that he, Chuang Tzu, had been dreaming that he was a butterfly. But then he asked himself a question: 'Was I Chuang Tzu dreaming I was a butterfly, or am I now a butterfly dreaming I am Chuang Tzu?' A nice question, but ultimately Chuang Tzu's dream was a purely human experience of 'being a butterfly': the real thing would have been too entirely alien to be comprehended; it would have seemed a sensory chaos.

Even in an age of science, butterflies are creatures of mystery and wonder. And the wonder they inspire is one of the ways in which we relate to these creatures so different from us, and one of the ways in which they speak to us of other worlds and other ways of being that we are only just beginning to understand.

The enchantment persists.

The Mindful Present: Seeing and Being

Looking back at the history of our changing relationship with butterflies, you could say that it has evolved through three ages, and is now in a fourth. First came the long age of indifference, in which butterflies were only casually noticed and given vague symbolic or moral meanings. Then came the aesthetic age, embodied by the Aurelians, in which an appreciation of the beauty of butterflies went hand in hand with the beginnings of scientific inquiry. The third age was the great age of collecting, in which netting, killing and pinning butterflies became a widely popular pursuit, while at the same time scientific knowledge made considerable strides. The advance of scientific understanding of butterflies continued, indeed accelerated, into the fourth age – the age of conservation, in which conserving what we have got, in the face of serious decline, is the primary concern. But this fourth age is also the age of watching, as against collecting – and of a particular kind of watching, which

I would call (in keeping with the spirit of the times) mindful watching.

That this is an age of watching, in the passive sense, is obvious: the characteristic activity of developed humanity today is staring at a screen, whether a computer screen, a television screen, or the ubiquitous mobile phone (no longer just a phone but a computer capable of holding, it seems, an entire life). Our relationship with the natural world has undoubtedly suffered as a result of this passive, screen-based relationship. As the naturalist Matthew Oates, quoted in Patrick Barkham's *The Butterfly Isles* (2010), remarked, 'To us, nature is something we do through the BBC Natural History Unit and a television screen. It scares me rigid what's going on. Our whole relationship is remote and not experiential.' Increasingly, the wildlife documentaries we watch are showcases for technology that is so advanced, so hyper-real, that it long ago ceased to bear much resemblance to the actual experience of observing nature in the wild. With ultra-high definition, ultra-long lens photography, extreme close-ups, slow-motion and time-lapses, infra-red imaging, the use of drones and micro-cameras, not to mention a certain amount of filming in constructed sets rather than in the wild, these shows are more gee-whiz spectacle than documentary.

The wildlife writer Richard Mabey, in his book *Nature Cure* (2005), writes feelingly of this, describing a time when, recovering from a depressive breakdown, he sought solace in watching wildlife documentaries. 'Sequences of carnivores pursuing game seemed like endlessly repeated film-loops, and caricatures of the complexities of life in the wild,' he recalls. 'Birds had miniature cameras strapped to their backs so that we could "share their view of the world" ... Almost every programme, however honourably intended, seemed bent on belittling the natural world, putting it firmly in its place.' Mabey wonders what the makers of these films think they are doing: 'Do they view the world much as the eighteenth-century makers of "cabinets of curiosities" did, as a collection of diversions and amusements, to be attractively presented behind glass? Do they really believe that the technological translation of the natural world – Slower! Closer! Bigger! – helps us understand how we fit into it?' Far too many people form their ideas of what nature is like, and what is happening to it, from watching these programmes, rather than from going out and experiencing it for themselves.

Many of those most vocal on the subject of Nature and the imminent threats to its survival have little real knowledge of it, and rarely look about them at what is actually there, let alone spend time in the open air observing and experiencing. They will be the first to tell you they never see butterflies any more, or wild flowers, or birds – but have they been looking? As the novelist Elizabeth Taylor observed of her monstrous creation, the romance writer 'Angel' (in the novel of that name), 'She was always too busy writing about what she thought of as "Nature" to go out of doors to look at things.' Those of us who do venture outdoors and look around us – especially those of us who are looking for butterflies – soon learn that the more we look, the more we will see; that nothing in nature is simple; and that there is little in it about which we can generalise.

Happily, as well as this passive, screen-based watching, there is another possibility: the active, mindful form of watching, in which we leave our screens behind and make a serious effort to open our eyes and minds (and all our senses) to what is around us, and, by doing so, inhabit the present moment. Mindfulness has been defined as 'the practice of being fully present'. It requires particular skills, a particular mindset, and it brings particular rewards for the mental and physical wellbeing that is being sought by so many, and in so many ways.

As with the conservation movement, the birders have got there first, with 'mindful birdwatching' now a recognised way of watching birds that is not concerned with ticking species off a list (the twitchers' obsession) but concentrates on the actual experience of watching and on the connection with nature that can be forged by letting go of our everyday concerns and being present in our experience of the moment. A leading advocate of this approach is Claire Thompson, who in 2017 published *The Art of Mindful Birdwatching: Reflections on Freedom and Being*, in which she explores 'what we can all learn from birds to welcome greater well-being into our lives'. Thompson sees birdwatching as 'the perfect entry point to rekindle our sense of what it's like to be truly human – and an integral part of the natural world'. There are now websites galore devoted to the mindful approach to nature in general, and there are birdwatchers who happily describe themselves as 'mindful birders'. Where birders go, surely butterfly watchers – mindful butterfliers? mindful aurelians? – can follow. Indeed, watching butterflies might also be described as 'the perfect entry point', and may even have more to offer the mindful watcher than birds …

What mindful butterfly watching offers and what it teaches might be summed up under three headings: perspective, attention and delight. We gain perspective from engaging with the natural world and acknowledging our place in it, as a tiny part of a vast scheme that spreads out around us in its infinite scope and variety; the close attention to and observation of what is around us works to draw us out of ourselves; and the particular delight that is to be found in watching butterflies can give us a taste even of timelessness, of being so completely immersed in the present that we seem, for that moment, actually to escape time.

Prisoners inside ourselves as we are for most of our waking life – caught up in the quotidian business of living, in our preoccupations, plans and desires – we inevitably make of ourselves something bigger and more central to the world than we really are. We loom large, we fill the frame; it is as if we have a telescope trained on ourselves all the time. We need to turn the telescope around, look through the 'wrong end', and get a better sense of our proper place in the scheme of things. Some theologians talk of the useful exercise of taking a 'God's-eye view' of ourselves and our situation (useful even if we don't ostensibly believe in God); what we might call a 'Nature's-eye view' is closely related. The perspective of Nature cuts us down to size and, in so doing, offers us a new way of being in the world. Watching butterflies offers a particularly effective way of achieving that perspective.

What is it about butterflies? What distinguishes butterfly watching from bird watching (apart from the fact that there are far fewer species, and you don't usually need binoculars to see them)? A key difference is that the season in which we can watch butterflies is quite short, restricted to the months between early spring and early autumn. This, combined with the relative paucity of butterfly life, the elusive ways of many species, and a general decline in abundance, makes the pursuit particularly intense, and its pleasures especially precious. In the course of the season, species come and go, some reappearing after their first emergence, others disappearing until the same time next year. The butterfly world is constantly in flux, more so than the bird world, in which resident species are to be seen all year round, and migrants throughout their particular season. With butterflies, each short season is punctuated by the excitement of first sightings of specimens freshly emerged and full of vigour, and the bittersweet sadness of seeing the last tatty, worn-out survivors. Butterfly lives are short; the window of opportunity for seeing some of them can be very brief indeed – especially if, as is too often the case, it is further shortened by adverse weather conditions. Butterfly watching offers particular challenges too, because, as well as being often elusive, butterflies are also relatively small and can be extremely mobile, as anyone who has chased after a fast flier will attest. Watching butterflies demands an intensity of attention and focus that is, I think, quite unique. We need to get near (a challenge in itself) and look closely at creatures that can be reluctant to reveal themselves to us, and might be gone in a flash before they have done so. To follow a butterfly in flight requires total concentration: shift your gaze for a split second, or even blink, and, when you look back, the chances are you will have lost it.

To get involved in watching butterflies is to enter a new world, one that is rich, vibrant, abundant with life and colour and energy – and in which we figure only as marginal, fleeting presences, potential threats but of no other interest. This parallel world goes on, with or without us. Butterflies had already been around for some fifty million years when the first specimens of *Homo sapiens* walked the earth, and another kind of butterfly altogether had evolved (from lacewing flies) and thrived for forty-five million years, then disappeared for another full forty-five million years, before *Lepidoptera* evolved. There are fossil remains of butterflies: one notably beautiful specimen, of an otherwise unknown species, is in the collection of the (London) Natural History Museum. Amazingly, its banded wing markings are still visible after thirty-four million years. It was discovered in the late nineteenth century, at Gurnard Bay on the north coast of what is now the Isle of Wight and was then a subtropical rain forest.

These dizzying perspectives of time are worth bearing in mind when we, latecomers to the carnival of life, look at butterflies. Though their individual lives are short, they are inhabitants of deep time, not of our transient moment. Those butterflies dancing in the air are not dancing around us, still less dancing to entertain us; they are just doing their thing, doing what they have always done, pursuing their *conatus*. As Spinoza would put it, they are persisting in their own being. We are outside their world. And yet, in the effort to enter it, to experience more of it as if from the inside, we have much to gain and to learn, not least about our place in nature. To shift perspective and see ourselves as marginal to their world, as being, like the butterflies themselves, just another element in the wide web of being, is perhaps the first and greatest lesson we can learn.

To immerse ourselves in this new realm, we need to become fully conscious of the wider natural world around us: of the movements of sun and shade so crucial to butterfly activity, of every nuance and change of wind and weather, of the rhythms of time and of the seasons, of different types of plant life and qualities of light, of geology, of the total environment we are inhabiting. The French word *terroir*, much used by connoisseurs of wine, is useful here, denoting as it does a particular combination of soil, topography, vegetation, weather and climate, along with still more local factors ('the south side of the vineyard'), that define a place. *Terroir* is a matter of deep emotional importance to the French, for whom it carries connotations of authenticity, particularity and home territory. To the butterfly lover, it is perhaps a more practical matter, but the pursuit of butterflies certainly requires a sensitivity to *terroir*, something that the artist and butterfly man David Measures calls 'environment sense'. As you begin to watch butterflies, you will become increasingly sensitive, especially to the presence or absence of sun. Butterflies, which are all sun lovers, dependent on its warmth, often respond almost instantly to a sudden burst of sunshine on a grey day. Some treetop-dwelling butterflies take to the air and dance in the late sun when all below is in shade and lower-living species are already roosting in the grass. Wherever you are, even in town or suburb, if you turn your eyes to where the sunlight falls, you might be rewarded by the sight of a butterfly or two.

I recently discovered a stretch of downland quite close to my home. I say 'discovered', but I had actually visited several times before, usually to enjoy the sight of a beautiful species, the Dark Green Fritillary, that flies there in its due season. It is not a very prepossessing site, compared with more picturesque areas of downland a little farther off, and it was to the latter that I tended to gravitate. But then along came Covid-19 and lockdown, and I found my horizons suddenly narrowed. I began to pay frequent visits to this more local patch; indeed, as the year went on, I began to haunt it. I got to know it much more intimately than before, and to wander around it for longer stretches, with closer attention, and at different times of the season. The result was that, as I became more fully acquainted with the micro-habitats that it contained – woodland edges, close-cropped turf, scrub and longer grass, shrubs and bushes, chalk scrapes, clumps of flowers attractive to particular species, sheltered patches and warm hollows, sun-facing slopes – I was rewarded by the discovery that this place was home to far more species of butterfly than I had ever imagined. I saw species ranging from lively little Grizzled Skippers, tiny Small Blues and emerald-winged Green Hairstreaks in the spring, to milky-blue Chalkhill Blues in late summer. I had immersed myself fully in this little world for the first time, and in doing so discovered that it was an environment far richer, more abundant and various than I had ever guessed. Instead of passing by, I had placed myself in the midst of it, and thereby discovered what I had been missing. And yet I was never at the centre of this world; my presence was of no significance. My experience of the place had made all the difference to me, but none at all to the life there, that will go on as it has always done, with or without me. Of all that was going on around me in that place and that season, I was the least important element. But of course it was not going on around me; it was just going on.

Since we humans are part of nature and not apart from it (whatever we might tell ourselves), we are taking our proper place when we engage with it in this immersive way. It is, in effect, a reversion to a childhood way of being in nature. Small children take it for granted that they are part of nature, that they and the animals around them are in this together, and part of the same world. This is why so many children's books are populated by animals who slip easily between their animal nature and a quasi-human identity, inhabiting both with no sense of incongruity. Seeing no discontinuity between the human and natural world, children are intrinsically 'interested in nature', but most lose that interest as they grow older. When David Attenborough was once asked 'When did you acquire your interest in nature?', he replied 'When did you lose yours?' Sadly, in these days of highly protective parenting, when children are no longer allowed to roam free as they once did, they have less chance of developing that interest at all. It is heartening though, that even in these straitened circumstances, so many still do – even if, as has always been the case, for many the interest does not last beyond childhood. It is also good to see the recent growth of 'forest schools' – by no means limited to forests – in which children are left free, with a degree of supervision and guidance, to explore whatever forms of wildlife are to be found.

It is hardly surprising that the more of nature we have in our lives – in physical form, not on a screen – the better our physical and mental health is likely to be. Indeed, this is so well documented that doctors are now prescribing walks and gentle exercise in the open air for mild depression, and scientific research has established the measurable benefits of such simple things as having vegetation around us (even indoors), or a window looking out onto something green, or sitting peacefully communing with our pets, those invaluable mediators between the natural world and the human.

There is now a large and ever-growing body of research demonstrating the benefits to health and wellbeing of contact with the natural world, with beneficial effects on everything from cardiovascular disease, diabetes, high blood pressure and obesity to mental health. The Japanese have been on to this for years, and their therapeutic practice of 'forest bathing' (***shinrin-yoku***) – cultivating calm and relaxation among trees – is now becoming popular in Britain too, under the auspices of Forestry England. Contact with nature, it has been found, reduces blood pressure and levels of the 'stress hormone' cortisol, while even the gentlest exercise outdoors increases the production of endorphins, the 'happiness hormones'. The very smells of the outdoors have been shown to have beneficial effects on mental and physical health. As for watching butterflies, research conducted by Butterfly Conservation in collaboration with the University of Derby in the wake of the 2022 Big Butterfly Count showed that spending just fifteen minutes tuning in to nature and counting butterflies reduced anxiety by an average of nine per cent, while enhancing general mental wellbeing. And these beneficial effects seemed to last for six or seven weeks after the Count itself. Really, we don't need scientists to tell us all this – we have always known it from our own experience and folk wisdom.

The kind of direct immersion in a particular natural environment that I've been describing is at the opposite extreme from the alienated, mediated, arm's-length (and then some) experience of nature that is increasingly the norm. When we are out there, immersed in it and open to what is around us, we discover that, once we start looking closely at it, nature is infinitely rich and various, and often unpredictable. Our butterflies, as well as exhibiting great variety in size, colour and markings across a relatively small number of species, display different characters and personalities, a wide range of behaviour, and often a high level of unpredictability. Though some butterflies are sedentary and set in their ways, others are much more flexible and are liable to turn up in the last place you'd think to look for them, quite disregarding the textbooks. To watch butterflies, even when they are behaving 'typically', we need flexibility and openness to adapt to their different ways of being and of expressing their nature, and to the fact that often they don't follow the rules. For them, after all, there are no rules.

'To see clearly,' wrote the Victorian sage John Ruskin, 'is poetry, philosophy, and religion – all in one.' And when he spoke of seeing clearly, he meant exactly that; he was not speaking figuratively. The relentlessly didactic Ruskin regarded the art of seeing as something that had to be taught: 'To be taught to write or to speak – but what is the use of speaking, if you have nothing to say? To be taught to think, if you have nothing to think of? But to be taught to see is to gain word and thought at once, and both true.' If Ruskin's Victorian readers had to be taught to see, how much more do we need it now, living as we do in a boundless ocean of visual and other distractions? It has never been harder than it is now, in our multi-media digital age, to see clearly, to focus our distracted, diffused perception on one thing, to pay attention.

Ruskin's perspective was that of an art critic and theorist, one who believed that the path to seeing clearly was through looking intently with the eye of an artist – or at least a sketcher, the best most of us can hope to be. In a long (too long to quote) passage in an 'Essay on the Relative Dignity of the Studies of Painting and Music' (1838), he contrasts the experiences of two people, one a sketcher, one not, taking a walk down a green lane. The non-sketcher will see a generic lane and green trees, sun and shade, an old woman in a red cloak, and will return home with nothing particular to report. The sketcher, on the other hand, will see ... Well, Ruskin describes what the sketcher sees at great length and in rhapsodic prose, leaving no doubt that the sketcher has had an infinitely richer experience, because he has exercised his trained powers of looking; he has seen clearly.

Though I would not make Ruskin's exalted claims on behalf of the butterfly watcher, there is an analogy to be drawn. A butterfly watcher, constantly on the lookout and using his powers of observation, will be more alert to and aware of the particularity of that green lane and what it might have to offer (Speckled Woods at the very least, by the sound of it) than his companion. He will indeed have the richer experience; he will have seen more deeply into what was there, what was around them both.

Despite his intense appreciation of beauty in nature and his love of a subject that would challenge his draughtsmanship, Ruskin was not particularly interested in butterflies, and seems only to have drawn a butterfly once (a not entirely convincing Amanda's Blue on an alpine thistle). However, no one was a more passionate advocate of the importance of looking intently and paying full, concentrated attention to what is before the eye. Such attention was, of course, key to drawing accurately from nature, but it was more than that – 'poetry, philosophy, and religion' indeed. For Ruskin, such intensely focused attention amounted to a moral imperative. In the following century, the French philosopher Simone Weil loaded the act of attention with still more moral significance: 'Attention, taken to its highest degree, is the same thing as prayer ... Absolutely unmixed attention is prayer.'

The novelist and philosopher Iris Murdoch, borrowing Weil's term but dropping the theology, defined attention as 'a just and loving gaze directed upon an individual reality ... the characteristic and proper mark of the moral agent.' She sees attentiveness to what is there, outside us, as a way of escaping the self and finding what she calls 'the unself'. In a passage in *The Sovereignty of Good*, Murdoch writes of looking out of her window in an anxious, brooding state of mind and suddenly seeing a hovering kestrel: 'In a moment everything is altered. The brooding self with its hurt vanity has disappeared. There is nothing now but kestrel. And when I return to thinking of the other matter it seems less important. And of course this is something which we may also do deliberately: give attention to nature in order to clear our minds of selfish care.' In order, that is, to 'take us out of ourselves'. Murdoch's kestrel could equally well have been a butterfly; it would have had just the same effect, if proper attention was focused on it: 'there is nothing now but butterfly'.

I quote Murdoch, Weil and Ruskin simply to illustrate how the act of paying close attention, in itself, carries greater significance than we might at first think. It does indeed have moral weight, if only because, when we pay such attention to something outside ourselves, we are paying that much less attention to ourselves; we are, to a greater or lesser extent, stepping outside ourselves and into a wider reality. And looking outside ourselves, paying attention to what is around us rather than navel-gazing, rather than walking around with our ears deafened to the outside world by earphones and our eyes fixed on the delusive light of our mobile phone screens, is demonstrably beneficial.

Like most of us, I don't set much store by research findings unless they confirm what I already believe to be true ('I may not know much about science, but I know what I like', as Martin Amis once remarked). I was pleased, therefore, to come across Dr Richard Wiseman's book *The Luck Factor* (2003), which looks into the differences between people who believe themselves to be lucky and those who believe themselves unlucky. When tested for their powers of observation, the 'lucky' types were found to be far more observant than the 'unlucky' ones. By attentiveness to what was happening around them, the 'lucky' ones had, in effect, created their own luck – spotting opportunities, possibilities, different ways of doing things, potential threats, unexpected developments – while the 'unlucky', by not paying enough attention to notice what was there, had reinforced their own perceived bad luck. So, pay attention and get lucky? Or just pay attention anyway; it is its own reward, especially if the objects of our attention are butterflies.

Close observation is an essential part of watching butterflies – observation of pattern and colour, flight and behaviour, all the factors that allow us to identify species and distinguish one from another. However, identification is not everything, and there is much to be said against the identifying tendency. Some advanced 'twitchers' (obsessive birdwatchers) seem to reach a point where they barely deign to notice anything but rarities – and they only notice them in order to tick them off their lists. The attention they pay to birds serves *only* the purpose of discriminating between species in order to find the ones of interest to them. It's a tendency I notice, in nascent form, in myself: it is all too easy to dismiss the more common species of butterfly, the mind merely registering 'just another Small White', 'just another Meadow Brown', even 'just another Speckled Wood' – and that, with its dappled brown-and-cream wings, is a butterfly of rare beauty that suffers from being too abundant to be properly appreciated. In other words, skill at identification can lead to *not* actually paying due attention to what you are looking at.

To pay full attention to the natural world – in particular, to butterflies – demands patience, perseverance and flexibility, among other things. One of the most important is the ability to stand still and wait. Standing still for any length of time is something that is surprisingly hard to do, but is well worth learning, because the rewards – both in terms of butterfly watching and general wellbeing – are considerable. To stand still is to calm down mind and body, to slow your systems to a pace at which you can open up to the natural world around you, and that world can open up to you. While you are in motion, all the observant life around you will be well aware of your presence, and much of it will take evasive action; when you are standing still, you become after a while something peripheral, of no particular interest, a minor feature in the landscape. After only a few minutes of standing still, you will find your surroundings gradually coming to life as more and more organisms resume their activities, barely noticing that you are there. Much of this wealth of active life will be in insect form, and only some of those insects will be butterflies, but by standing still you have the best chance possible of observing them going about their business, unthreatened by your alien presence.

In his delightful book *Bright Wings of Summer* (1976), David Measures writes eloquently of the rewards of standing still, and also describes 'something extra' that can take place in such intense encounters with butterflies: how, 'after a period of watching, your particular butterfly character appears to come to terms with you. Reconciled to your presence, it seems to allow a trust to exist, whereby both of you take part, each functioning in your own way, freely and co-existent.' At such moments, a kind of communion seems to develop between observer and observed, and the act of butterfly watching ceases, for a moment, to be entirely a one-way process; subject is also object, and object also subject. By standing still, we have allowed something extraordinary to happen.

Attentiveness is a wider and more inclusive thing than the fixed attention that we direct at the butterfly itself, as it also involves attention to the environment as a whole, to the totality of the *terroir*. In the process of watching butterflies, we become sensitised to the aspects of a particular environment that might make it attractive to butterflies, or to particular kinds of butterflies. This is a broad and salutary attentiveness, an openness to the wider picture, something that goes far beyond the identification and ticking-off of species.

In the foreword to *Bright Wings of Summer*, Sir Peter Scott writes of 'the aesthetic beauty, scientific interest and philosophic delight of studying living butterflies'. This is a nice formulation, and I especially like the phrase 'philosophic delight', which points to an aspect of butterfly watching that tends to get passed over in most books on the subject: the pleasurable stimulation of thought and imagination that the sight of these creatures can generate. The pursuit also brings with it other forms of delight, including something more immediate and straightforwardly physical. Vladimir Nabokov wrote of experiencing a spine-tingling physical pleasure in the pursuit of butterflies which he found comparable only to the sensation he felt in the presence of great writing or great art. 'The spine knows best,' he wrote. 'The seat of artistic delight is between the shoulder blades.' That 'little shiver behind', that sudden *frisson*, was for him the index of true art and true emotion, and he experienced it alike in the presence of man-made beauty and of the natural beauty of butterflies. Nabokov was as much a scientist as an aesthete when it came to butterflies, so his pleasure had a double nature. He claimed that 'I cannot separate the aesthetic pleasure of seeing a butterfly and the scientific pleasure of knowing what it is.' Clearly he felt both as essentially the same thing, envisaging a 'high ridge where the mountainside of "scientific knowledge" joins the opposite slope of "artistic imagination"'.

When Alfred Russel Wallace, the great naturalist (who, independently of Darwin, realised that evolution could work through a process of natural selection), had his first sighting of the magnificent butterfly that was to be named after him – Wallace's Golden Birdwing – he reported that 'my heart began to beat violently, the blood rushed to my head, and I felt much more like fainting than I have done when in apprehension of immediate death [which was for the adventurous Wallace a frequent experience]. I had a headache for the rest of the day.' Wallace's reaction is extreme, but a less dramatic (and certainly less headache-inducing) version of it is something I have often experienced myself, on coming across some scarce or elusive butterfly, or chancing unexpectedly on a perfect bit of habitat rich in butterfly life.

This thrill of the unexpected is something I felt more than once during that lockdown period: when I first encountered Grizzled Skippers and Chalkhill Blues on the rediscovered downland site I wrote about above; when I found Green Hairstreaks flying at another nearby site where I had never seen them before, nor expected to see them; when a perfect pair of spectacular Silver-Washed Fritillaries made a one-off appearance in an unlikely location; and when I had a late-season run of three wholly unexpected sightings of the elusive Brown Hairstreak. With each of these encounters I certainly felt something of Nabokov's 'little shiver behind', and I don't doubt that my heartbeat sped up and my brain was flooded with joyful endorphins. I know that at such moments I habitually catch my breath and gasp. In fact I experience something that the American poet Kay Ryan identifies in her own reaction to reading great poetry – 'a kind of giddiness indistinguishable from the impulse to laugh'. Ryan likens the reaction to 'one of those involuntary *ha!s* that jump out when you've witnessed a wonderful magic trick'. I am much given to those *ha!s* when I'm butterfly watching, and to a kind of short, quiet laugh that can leave me smiling idiotically for some while afterwards.

I also, I must admit, talk to myself quite a lot (when I'm sure I'm on my own – I don't want to appear too much the mad butterfly man) and often address the butterflies, gently urging them to come forward and show themselves, or to stop darting about and settle somewhere I can take a good look at them. 'You beauty!', I often exclaim, *sotto voce*, when one unexpectedly lands close by – or even on me, as quite often happens – and strikes a perfect pose. The beautiful surprises that the butterfly world presents us with can seem very like some wonderful kind of magic. The pleasure they give to the receptive is an intense, heart-lifting delight, perhaps the richest reward of the butterfly-watching life.

The delight of butterfly watching lies partly in the pleasure of pursuit and partly in the beauty of the quarry when (if) we find it. It is particularly intense because, if we are not netting it or photographing it, we know we must enjoy that butterfly to the full while it is there, so briefly, in front of us; it could be gone any second, likely never to be seen again. The delight of spotting it is a transitory and precious pleasure, of which a large element is aesthetic: our butterflies, even the least spectacular of them, are objects of beauty. Perhaps I should confess at this point, what is no doubt already obvious, that my own interest in them is in essence more aesthetic than scientific. It is the beauty of their wings, above all, that draws me to them and gives me that special delight. In this I am closer in spirit to those early aurelians than to the scientific lepidopterists of more recent times.

Aesthetic delight and the 'little shiver behind' might be classed as transient thrills, as flashes of joy, but there is also a steadier, calmer, more continuous pleasure that goes with butterfly watching, a glow of background delight compounded of various elements

– one of which is, simply, walking. There is a Latin saying, *Solvitur ambulando* (it is solved by walking), which originally had a philosophical meaning referring to problems that can be solved by practical experience or demonstration. Diogenes the Cynic is said to have applied it literally, refuting Zeno's paradoxes about the impossibility of motion by getting up and walking away (much as, many centuries later, Samuel Johnson addressed Bishop Berkeley's theory of the immateriality of objects by kicking a large stone and declaring, 'I refute it thus!'). Over time, *solvitur ambulando* took on a wider meaning, one more related to the beneficial effects of walking, an activity that can indeed play a useful part in solving problems, if only by stimulating thought (I'll be returning to that theme shortly). The phrase got a new lease of life when the travel writer and novelist Bruce Chatwin used it in *The Songlines* (1987), having picked it up from his friend and mentor Patrick Leigh Fermor. Fermor was a heroic walker (and self-mythologiser) who in his youth had travelled on foot from the Hook of Holland to Istanbul, and Chatwin was a man who 'passionately believed that walking constituted the sovereign remedy for every mental travail'. When he heard Fermor use the Latin phrase, it went straight into his notebook, and re-emerged in *The Songlines*, a book about the 'songlines' or 'dreaming-paths' of aboriginal Australians that celebrates the deeply human, richly satisfying activity of walking about.

Butterfly watching is, in essence, a form of walking – not, like golf, 'a good walk spoiled', but rather a good walk greatly enhanced by the chance of seeing some very beautiful creatures. Walking brings with it all the pleasures and benefits of being in the open air and in a steady, easy form of motion. Unlike running, walking takes us slowly, at a human pace, through the landscape, with time to take in the sights and sounds and scents of nature, to enjoy them all and reap the benefits. I need hardly add that the kind of walking I am talking about here is not the Serious Activity involving specialist kit, objectives and challenges, that drives some across the landscape – no, it is more like a form of sauntering.

The American sage Henry David Thoreau, in his essay/lecture on 'Walking', declares that 'I have met but one or two persons in my life who understood the art of Walking, that is, of taking walks – who had a genius, so to speak, for *sauntering* ...'. Thoreau, one of the greatest, most observant and receptive of walkers, commends sauntering as the ideal form of walking. Watching butterflies is a kind of focused sauntering: walking with a purpose, yes, but one that is fluid and unpredictable, that might lead us anywhere or nowhere, and will abide no rigid programme. It is essentially walking for pleasure, but it brings with it tangible benefits.

There is no better way of thinking through a problem than taking a walk, for the rhythms of walking are the rhythms of thought. I'm sure I am not the only writer who gets most of his better ideas, and composes most of his better passages, when walking: I even had the idea for this book while walking – among butterflies, needless to say. Walking – especially walking alone – offers a kind of suspensive freedom in which the mind can operate more freely than when hemmed in by circumstance and moving to rhythms imposed by necessity. Nietzsche, who was a prodigious walker, believed that 'All truly great thoughts are conceived while walking', and once wrote an entire book (*The Wanderer and His Shadow*, 1880) from notes made while on his epic walks. Never one to understate his case, Nietzsche counselled his readers to 'give credence to no thought that is not born in the open air and accompanied by free movement', and even declared that sitting still was 'the real sin against the Holy Ghost'.

Kierkegaard, a philosopher of a less fire-breathing disposition than Nietzsche, wrote in a letter that 'every day I walk myself into a state of well-being and walk away from every illness; I have walked myself into my best thoughts, and I know of no thought so burdensome that one cannot walk away from it ... but by sitting still, and the more one sits still, the closer one comes to feeling ill ... Thus if one just keeps on walking, everything will be all right.' *Solvitur ambulando* indeed.

Charles Darwin every day walked along a sand-covered path winding through woods and along a field edge adjoining the grounds of his house at Downe in Kent. Darwin called this 'my thinking path' and he would think through scientific problems as he walked. Sometimes he would stack stones at the start of the path, knocking one away each time he completed a circuit. Like Sherlock Holmes with his 'three-pipe problems', Darwin would anticipate a 'three-flint problem', heading home when three stones were gone. Though it would be fanciful to suggest that we owe *The Origin of Species* to Darwin's daily walk, surely much else fermented profitably in his brain on those circuits of the sand-covered path.

But we have sauntered a long way from our subject. To return to the delights of butterfly watching, two that are particularly intense are the sighting of the first butterfly of the year and the last. You can never be certain about the latter – will that faded Red Admiral on a late ivy flower really be the last, or will I see another? The final few sightings of the year have a poignant flavour. The summer is all behind, the winter all ahead, its cold dark months devoid of butterfly activity. Usually I know that I have seen my last by the end of November, but still I live in hope of a late-season surprise, or a hibernator deceived into waking by a little unseasonal warmth (in 2020 I saw my last, a Red Admiral, in mid-December, and my first of the next season only seven weeks later).

The wait for the first butterfly of the new season begins as soon as there are signs of spring, and the keen pleasure of anticipation is followed in due course by the sheer joy of the first sighting – a joy matched only, in my experience, by the first Swift of summer. The first butterfly might be a newly awakened male Brimstone, sulphur yellow against a background of dark ivy, patrolling a woodland edge or railway embankment; or a basking Peacock soaking up the weak sunlight on a patch of bare ground; or a Comma on an ivy leaf, or a Tortoiseshell flying weakly past. The one thing we can be sure of is that we never know exactly when it will be – that soul-stirring moment when the first butterfly of the year presents itself to us, giving promise of all the lepidopteral delights coming our way, or eluding us, as the season unfolds. So short is the butterfly year – less than two-thirds of the calendar year – and so short the flying season of some species that these pleasures must be caught, literally, on the wing. In the course of a season, we shall be having an exhilarating succession of first and (unknowing) last sightings, of greetings and farewells. For those of us whose pleasure in butterflies goes deeper than list-ticking, it is an intense business, replete with moments of sheer delight. Into this delight flow all the pleasures attendant on walking outdoors, surrounded by natural beauty and breathing fresh air; the aesthetic pleasure afforded us by the beauty of the butterflies; the pleasure of close attention (and determined pursuit) rewarded; and the pleasure of inquiry, of finding out more, as these fascinating and still mysterious creatures reveal something of themselves.

By becoming absorbed in this transporting delight, we can at times achieve the profound happiness that comes from losing ourselves in the moment. To live in the moment – the goal of mindfulness, and of all who seek a sense of being fully alive – is, the philosopher Wittgenstein suggests, to taste the eternal: 'If we take eternity to mean not endless temporal duration but timelessness, then he lives eternally who lives in the present.' No one has written more eloquently of experiencing this sense of timelessness when among butterflies than Vladimir Nabokov. At the end of the chapter of his autobiographical *Speak, Memory,* devoted to his butterfly life, he writes, 'the highest enjoyment of timelessness – in a landscape selected at random – is when I stand among rare butterflies and their food plants. This is ecstasy, and behind the ecstasy is something else, which is hard to explain. It is like a momentary vacuum into which rushes all that I love. A sense of oneness with sun and stone. A thrill of gratitude to whom it may concern – to the contrapuntal genius of human fate or to tender ghosts humouring a lucky mortal.'

This is delight indeed – and I will only add that the butterflies need not be rare (and you certainly don't need a net in your hand). Nabokov had a special love for that large tribe of small butterflies we call the Blues, and, reading that passage, I always envisage him in an alpine meadow surrounded by Blues dancing from flower to flower. I have myself experienced something very like the sensation he describes when standing enraptured on a Surrey hillside on a sunny summer day while Adonis Blues, their wings the colour of heaven, fly all around me. And I am lost in wonder and delight.

Part III

Flights in the Mind

The Parallel History: Dreams, Art, Music, Literature – and Obsession

The inward history of butterflies, in the mind and the imagination, in our dreams, and as embodied in poetry, music and art, is not often explored, and yet it is, I think, a vitally important, enriching element of the very special relationship we have with them.

Dreams, then. I have had butterfly dreams from time to time for most of my life. Sometimes I dream of coming across elusive rarities like the Camberwell Beauty, often in unlikely settings or at unlikely times of year, always to thrilling, spirit-lifting effect. I dream too of 'ordinary' butterflies, particularly of the Blues and the 'Aristocrats' – Emperors, Admirals, Vanessids and Fritillaries. Sometimes a single butterfly turns up to delight me, sometimes I am surrounded by beauties of all kinds, some of them clearly exotics, in a kind of butterfly paradise. Quite often, flying among them is a specimen I cannot identify, though I feel I know it, a creature that teasingly refuses to be named: if it isn't among the British butterflies, it looks as if it ought to be – but what is it? I suppose that when someone is so steeped in the love of butterflies as I am, it is hardly surprising that they would dream of them, but it seems that butterflies, perhaps more generically conceived, take flight in many people's dreams.

Butterflies certainly figure large in the books and online oracles that purport to interpret the meanings of dreams. These will tell you what it 'means' to dream of a butterfly landing on you, to dream of a swarm of butterflies, of a butterfly with a broken wing, of blue, black or white butterflies, even of the Monarch specifically. There are Biblical interpretations on offer, somewhat undermined by a frank admission that 'butterflies are not expressly mentioned in Scripture', and Islamic interpretations (which seem, like the Church of old, to take rather a dim view of butterflies). All of this suggests that there are plenty of people dreaming of butterflies. With their beauty, their wayward flight and their particular fragile and fleeting presence, these creatures untethered to Earth seem made to inhabit our dream world, to flit easily over the threshold of waking reality into a more intimate realm.

It seems that butterflies were present in the human imagination from earliest times: recognisable butterflies appear in Pyrenean cave art, in Egyptian tomb paintings and bas-reliefs, in ancient Mesoamerican and Aztec art, in Minoan artefacts, and of course in classical Greek art. These butterflies are in part symbolic presences – and not always benign. In Aztec symbolism butterflies were associated with fire and warfare and the spirits of dead warriors. The butterflies depicted in ancient Egyptian art seem to carry the least symbolic weight, being simply one element among the many familiar and agreeable things of this world that the dead are taking with them into the next life. Elsewhere, the potent symbolism of butterflies as liberated human souls is no doubt at work, most expressly in the art of ancient Greece. In the Christian world, this association is not very strongly represented in the visual arts. We can only guess at what Hieronymus Bosch meant when he gave two of the devils in his *Garden of Earthly Delights* (1490–1510) the wings of butterflies. One is a correctly painted Meadow Brown, the other a far from correct Small Tortoiseshell: in his novel *Ada* (1969) Nabokov conjectures that Bosch 'evidently found a wing or two in the corner cobweb of his casement and showed the prettier upper surface in depicting his incorrectly folded insect'. Similarly, it is hard to imagine what Pieter Breugel the elder was thinking of when he gave one of the angels in his *Fall of the Rebel Angels* (1562) the wings, considerably stylised, of a Swallowtail. In a strange allegorical painting by Dosso Dossi, *Jupiter, Mercury and Virtue* (1524), executed in sixteenth-century Ferrara, Jupiter is seen sitting at an easel, painting butterflies and thereby giving them life (though they look far from lifelike). In later art, the private symbolism of both Odilon Redon and Salvador Dalí featured butterflies, but these are surely butterflies of the dream world.

For something more naturalistic and more identifiable, the place to look is in the flower paintings that were so popular in the Netherlands in the seventeenth and eighteenth centuries. In the works of, among others, Maria van Oosterwyck, Adriaen Coorte and Elias van der Broeck, recognisable, correctly drawn butterflies often feature amid the floral splendour, enhancing the visual pleasure the good Dutch burghers derived from such exuberant compositions, while also providing a little *frisson* of symbolic meaning: all this floral beauty is frail and short-lived, like the butterfly – and look out for the caterpillars, those devouring, corrupting worms …

Perhaps the most touching and brilliantly executed of all butterfly paintings – or rather paintings with butterflies in them – is Thomas Gainsborough's *The Painter's Daughters Chasing a Butterfly* (1756), which shows the two little girls in summer dresses, taking a woodland walk. The elder, Mary, protectively holds her sister's hand, yet seems detached from the little drama that is about to play out as Margaret impulsively reaches out to grab a white butterfly that has perched on a thistle – a double jeopardy. Clearly the butterfly serves as an emblem of the transience of life and beauty and the brevity of childhood, while the thistle represents danger. Like most painters, Gainsborough was no lepidopterist – this White seems to have its upperwings where its underwings should be – but it is a brilliantly effective painting.

Nabokov takes a severe line on the failure of most artists, even the best of them, to paint butterflies at all accurately. 'Only myopia,' he declares (in a 1970 interview), 'condones the blurry generalisations of ignorance. In high art and pure science detail is everything.' At the time of the interview he has been researching butterflies in art, and 'one simple conclusion I have come to is that no matter how precise an Old Master's brush can be it cannot vie in artistic magic with some of the coloured plates drawn by the illustrators of certain scientific works in the nineteenth century [and the eighteenth, surely?]. An Old Master did not know that in different species the venation is different, and never bothered to examine its structure. It is like painting a hand without knowing anything about its bones, or indeed without suspecting it has any.' This is fair enough: we rightly expect a naturalistic artist to know what they are painting, to have looked long and hard, to have paid due attention to it. When it comes to butterflies, it is the illustrators who have generally done that far more effectively than the fine artists; it is their job to get it right. And in doing so, some – Maria Sybilla Merian, Moses Harris, Benjamin Wilkes, Frohawk, Richard Lewington – have created images that can be justly seen as works of art in their own right, compositions that beautifully ***and*** accurately express the essence of a butterfly and the whole cycle of its life: 'pure science' meets 'high art'.

Butterflies have been rather better served by music than by the visual arts. The dancing, gliding and dipping movements of butterfly flight lend themselves naturally to musical expression, and, with the development of the piano in the Romantic era, the butterfly became a particularly popular subject. With its wide tonal range and endless dynamic possibilities, no other instrument can so delicately and fluently evoke the butterfly's flight, as is brilliantly demonstrated by such solo piano pieces as Grieg's *Schmetterling*, and Moritz Rosenthal's *Papillons*. Schubert ('*Der*

Schmetterling'), Fauré ('*Le Papillon at La Fleur*') and Debussy ('*Les Papillons*'), among others, wrote piano-accompanied butterfly songs, but perhaps the most dazzling demonstration of the piano's ability to evoke the restless dancing movements of a butterfly is Chopin's 'Butterfly' Etude, opus 25, number nine. With its high-speed alternation of *staccato* and *marcato*, this works its magic both aurally and visually, as the hands of the pianist actually mimic the movements of butterflies flitting from flower to flower. String instruments, also, can effectively evoke butterfly flight, and a notable example is Fauré's *Papillon* for cello and piano, a virtuoso piece that was a favourite encore of Pablo Casals. More recently, the Soviet-born Australian composer Elena Kats-Chernin wrote *Butterflying*, a delicately evocative piano piece in the Romantic tradition.

Butterflies in literature are surprisingly few. There has been much writing about butterflies, of course, but little of it could be classed as literature. Nabokov, the most literary of all butterfly lovers, was conscious of the inadequacy of his own representations of butterflies in his fiction, when compared to his scientific work. In a 1969 interview, he declared that 'whenever I allude to butterflies in my novels, no matter how diligently I rework the stuff, it remains pale and false and does not really express what I want it to express – what, indeed, it can only express in the special scientific terms of my entomological papers. The butterfly that lives for ever on its type-labelled pin and in its O. D. (original description) in a scientific journal dies a messy death in the fumes of the arty gush.' 'Arty gush' is hardly a fitting description of Nabokov's own butterfly allusions, but he is surely on to something in his sense that the depiction of butterflies in literature – when they are present at all – lacks vividness and precision.

Shakespeare, as we have noted before, seems to have had a butterfly-shaped blind spot, and chasing butterflies through later periods of English poetry brings few rewards.

Benjamin Wilkes's *English Moths and Butterflies* (1749), described in the chapter on the Aurelians, is prefaced by a charming poem by Henry Baker, his friend and fellow Aurelian. It is not great literature, but is worth quoting for its heartfelt expression of Aurelian enchantment (and faith in God's instructive ordering of His creation):

> See, to the Sun the Butterfly displays
> His glistering Wings and wantons in his Rays:
> In Life exulting o'er the Meadow flies,
> Sips from each Flow'r and breathes the vernal Skies.
> His splendid Plumes, in graceful order show
> The various glories of the painted Bow.

Of the Romantics, Wordsworth wrote a short, exclamatory poem, 'To a Butterfly' (1807), in which the butterfly obligingly turns the poet's thoughts to his favourite subject – himself: 'Much converse do I find in Thee,/Historian of my infancy!' The butterfly recalls to Wordsworth 'the time, when in our childish plays/My sister Emmeline and I/Together chased the Butterfly!' The sister is of course Dorothy, and the butterfly is entirely generic.

The 'peasant poet' John Clare's 'To the Butterfly' (1821) approaches closer to the specifics of reality, describing a butterfly perhaps waking from hibernation in a barn:

> Lovely insect, haste away,
> Greet once more the sunny day;
> Leave, O leave the murky barn,
> Ere trapping spiders thee discern;
> Soon as seen, they will beset
> Thy golden wings with filmy net,
> Then all in vain to set thee free,
> Hopes all lost for liberty ...

'Golden wings' suggests a Brimstone, but no – later in the poem the butterfly reappears, still golden but decked in 'crimson, blue and watery green,/ Mix'd with azure shade between'. That sounds more like a conflation of beauties than any particular species, but at least Clare's poem shows evidence of real observation and real delight. Clearly Clare, who saw his boyhood haunts lost to land enclosure and ended his days in a lunatic asylum, identifies with the butterfly threatened with the loss of its liberty, but the poem is not, unlike Wordsworth's, all about him.

In his sonnet 'To an Early Butterfly' (1821) Clare begins with a lively description of the butterfly itself (again unidentifiable) but is soon more concerned with the religious meaning to be derived from its early appearance:

Thrice welcome here again, thou flutt'ring thing,
That gaily seek'st about the opening flower,
And opest and shutt'st thy gaudy-spangled wing
Upon its bosom in the sunny hour;
Fond grateful thoughts from thy appearance spring:
To see thee, Fly, warms me once more to sing
His universal care who hapt thee down,
And did thy winter-dwelling please to give.
That Being's smiles on me dampt winter's frown,
And snatch'd me from the storm, and bade me live.
And now again the welcome season's come,
'Tis thine and mine, in nature's grateful pride,
To thank that God who snatch'd us from the tomb,
And stood our prop, when all gave way beside.

Butterflies seem to have fascinated the New England poet Emily Dickinson and are a vivid presence in many of her short, strange poems. There is even something butterfly-like in the manner of her verse, which has a delicate, dancing quality, taking off in short, darting flights, then landing briefly – or not landing, in the way of a butterfly rising and falling amid grasses, and never quite settling. In 'The Butterfly Obtains' (1914), she writes with affectionate regard, querying the puritan moral disapproval of the feckless butterfly –

The butterfly obtains
But little sympathy
Though favourably mentioned
In Entomology –
Because he travels freely
And wears a proper coat
The circumspect are certain
That he is dissolute –

> Had he the homely scutcheon
> Of modest Industry
> 'T were fitter certifying
> For Immortality –

For Dickinson, the reclusive Belle of Amherst, the butterfly represents an enviable freedom and lightness:

> The Butterfly upon the Sky,
> That doesn't know its Name
> And hasn't any tax to pay
> And hasn't any Home
> Is just as high as you and I,
> And higher, I believe,
> So soar away and never sigh
> And that's the way to grieve –

Perhaps Dickinson's best-known butterfly poem is 'Two Butterflies Went out at Noon' (1914), which wistfully observes two butterflies taking off for who knows where – no earthly destination:

> Two Butterflies went out at Noon –
> And waltzed above a Farm –
> Then stepped straight through the Firmament
> And rested on a Beam –
>
> And then – together bore away
> Upon a shining Sea –
> Though never yet, in any Port –
> Their coming mentioned – be –
>
> If spoken by the distant Bird –
> If met in Ether Sea
> By Frigate, or by Merchantman –
> No notice – was – to me –

The English poet Edward Thomas, who died in the First World War, describes a memorable butterfly encounter in his poem 'The Brook' (1918). The (unidentified) butterfly flies down and lands near the poet while he is watching children at play in a stream, and man and butterfly briefly commune:

> And down upon the dome
> Of the stone the cart-horse kicks against so oft
> A butterfly alighted. From aloft
> He took the heat of the sun, and from below.
> On the hot stone he perched contented so,
> As if never a cart would pass again
> That way; as if I were the last of men
> And he the first of insects to have earth
> And sun together and to know their worth.

Incidentally, Thomas is exactly right about how butterflies take heat from both above and below.

It is a pity that there are not more butterfly poems, but most literary writers seem to be more interested in the strange humans who chase and collect butterflies. Even Nabokov left a notably unsympathetic portrait of an obsessive collector in a short story called 'The Aurelian' (1916). In this, an impoverished and unhappy butterfly dealer becomes so consumed by his dreams of making a grand collecting trip abroad that he swindles a customer and abandons his wife and business – or, rather, he is about to do so when fate intervenes...

Butterflies in poetry are typically seen as embodying an enviable freedom to which we earthbound humans cannot aspire. They are escaping, flying away, slipping the surly bonds of Earth. Or, like Wordsworth's butterfly, they remind us of something we have lost – our childhood, our past. Such nostalgia is certainly an element in my own love of watching butterflies; they are a link with other passages and moments in my life, with particular times and places. Often the butterfly memory is stronger than the memory of the place: Nabokov (later in the 1969 interview quoted above) gives an example of this phenomenon: 'if I hear or read the words "Alp Grum, Engadine" the normal observer within me may force me to imagine the belvedere of a tiny hotel on its two-thousand-metre-tall perch and mowers working along a path that winds down to a toy railway; but what I see first of all and above all is the Yellow-Banded Ringlet settled with folded wings on the flower that those damned scythes are about to behead.' That scene has the sharpness and dizzying perspective of an image from a dream.

It also sometimes seems that butterflies are 'there for us' in a quite unique way, as if they are messengers from elsewhere – the land of dreams perhaps, which they seem so easily to inhabit? Or even from the world of the dead? It is not hard to understand how the folk belief in butterflies as liberated souls took root. I remember one occasion when, on a cold November day just before my mother's funeral, a butterfly suddenly appeared in a most unlikely town-centre setting and flew towards me. It was – of course! – a Red Admiral, and it seemed to be a sign and a wonder. On another chilly November day, I stepped out of the house, having just had the news that my third grandson had been born – and there was a Red Admiral, happily nectaring on a late Viburnum. Butterflies, like no other comparable creatures, engage our emotions; they are *personal* to us, they seem to carry particular meanings and intimations – perhaps that, ultimately, all might be well, that our bodies are not the whole of us, that freedom and escape are possible, that our dreams might have wings.

No doubt it is because of this kind of emotional engagement that a fascination with butterflies can become so intense that it tips over into something rather less healthy – into obsession, as portrayed in 'The Aurelian', a short story by Nabokov. The annals of lepidoptery are rich in obsessive collectors who seemed unable to stop, however unmanageably vast their collections became. We have come across one already – Percy Bright, the Chalkhill Blue obsessive (see 'My Butterfly Life') – but there were many others every bit as obsessed as him. Herman Strecker, a sculptor and stone carver from Philadelphia, was driven by a burning desire to possess at least one specimen of every species of Lepidoptera on the planet. He was a strange, eccentric and difficult man who was suspected by some of purloining specimens from other people's collections and even the American Museum of Natural History (though nothing was ever proved). What pleasure he got in acquiring his specimens seems to have been more than matched by the pain of longing to possess species that eluded his grasp: 'Why did God implant in us unquenchable desires, and then deny the means of gratifying them?' he lamented in a letter, sounding more like a frustrated lover than a lepidopterist. He was, however, a serious lepidopterist, who published important works on the subject, illustrated by himself (he was a talented artist). Strecker spent the last four decades of his life, and of the nineteenth century, amassing a collection of around two hundred thousand specimens, and identifying some three hundred new species. This collection, the largest private collection in the New World, occupied a whole floor of his Reading (Pennsylvania) town house, and it was purchased after his death by the Field Museum of Natural History of Chicago.

Strecker's collection, however, was as nothing compared to that of (Lionel) Walter Rothschild, 2nd Baron Rothschild. This scion of the great banking dynasty was not cut out for the business world, and devoted most of his time and energy to his vast private museum at Tring in Hertfordshire, which grew to become, in the words of his niece Miriam (herself a distinguished entomologist), 'the greatest collection of animals ever assembled by one man'. It included 300,000 bird skins, 20,000 birds' eggs, 144 giant tortoises, 300,000 beetles – and some two and a half million mounted butterflies and moths, representing 100,000 species of butterflies and larger moths and a vast range of variants. All of these collections were serious scientific resources, on the basis of which Rothschild and his collaborators described 5,000 new species and published some one thousand two hundred books and papers. A huge man, with surprisingly small feet, Rothschild seemed to move about his museum 'like a grand piano on castors' (in Miriam's phrase). He was extremely shy, but had a powerful, unpredictable presence, and was given to eccentric behaviour, as when he harnessed some of his zebras to his carriage and drove along Piccadilly to Buckingham Palace, or when he rode on the backs of his Aldabran giant tortoises, urging them on with a lettuce on the end of a stick. When he died in 1937, his collections and library (30,000 volumes) were left to the nation, as was his museum at Tring, which is now an outpost of the Natural History Museum, whose London collection, and its scientific work, were transformed by Rothschild's unparalleled bequest. He may have been an obsessive collector, but the fruits of his obsession served, and still serve, science.

The same could not be said of the vast natural history collections of the eccentric Sir Vauncey Harpur Crewe, which included huge numbers of exotic butterflies and moths, a hoard that, at its peak, was said to be second only to Rothschild's in size. Sir Vauncey was so fixated on his precious collections that he ordered fires to be maintained day and night in every room of Calke Abbey, the family seat in Derbyshire, to maintain ideal conditions. Any servant who failed in this duty was issued with an immediate dismissal notice, but as Sir Vauncey hardly knew one servant from another, these were usually ignored. One of his staff, however, was his constant companion – the head gamekeeper, the gloriously named Agathos Pegg, with whom Sir Vauncey would go on collecting expeditions on his extensive estates. Once, when the unlikely pair were collecting butterflies near Repton Park, a house occupied by one of the Harpur Crewe cousins, there was an altercation of some kind, following which Sir Vauncey ordered that Repton Park be razed to the ground, which it duly was. The bulk of Sir Vauncey's collection of Lepidoptera was sold after his death, in a total of 412 lots, but happily some of it remained at Calke Abbey, along with large numbers of stuffed and mounted specimens, trophy heads, birds' eggs and all manner of miscellaneous objects accumulated by this truly obsessive collector. The house is now in the hands of the National Trust, who, rather than tidying it up, have commendably preserved it as a portrait of 'the dramatic decline of a country estate', complete with peeling plasterwork, abandoned rooms – and all that mad, magnificent clutter. Sir Vauncey was undoubtedly a man in the grip of obsession rather than a seeker after knowledge, or even very much of an expert (he was reputed to be a soft touch for dealers with dubious specimens to sell).

Lacking the vast resources of a Rothschild or a Harpur Crewe, most obsessives have to specialise, often in a single species. A surprising example of a single-species obsessive was Joseph 'Joey' Grimaldi, the actor, comedian and dancer who was a huge star of the Regency stage. His obsession was with the 'Dartford Blue', the exquisite butterfly we now call the Adonis Blue, and it seems to have been born of a kind of aesthetic mania: he simply could not get enough of that heavenly blue. Like Dickens (who edited, or rather wrote, his *Memoirs* in 1838), Grimaldi was a man of phenomenal energy. In the summer, he would habitually leave London after his evening performance at Sadler's Wells, walk the fifteen miles to Dartford, spend the morning chasing his beloved Blues, then walk all the way back to London in time for his evening performance. He was quite capable of keeping this up for two or three nights in a row. At his home in Pentonville, Grimaldi amassed a collection of perhaps four thousand butterflies, most of them Adonis Blues, but it was lost when thieves broke into his house and, seeing no value in the collection, smashed up the cabinets that contained it, leaving nothing intact but one small box and a few items of collecting kit, which Grimaldi gave away to an acquaintance the next day. He never again took up his net, but in his sad later years, as he declined into ill health and alcoholism, he would look back fondly on the days he spent chasing the Dartford Blues.

One British butterfly species has inspired more obsessive attention than any other – the Purple Emperor. The reasons are not far to seek, as it is a large and spectacular beast, elusive and hard to find, eccentric and unpredictable in its habits, and it seems to invite a kind of personal relationship – now teasing, now imperious, now downright aggressive – with those who pursue it. No wonder the Emperor is known to his devotees as 'His Imperial Majesty' or 'H.I.M.'. In the days of butterfly collecting, the Purple Emperor was prized above almost all British butterflies, representing the ultimate challenge to the collector. The Emperor habitually lives high up among the treetops, is a formidably strong flier, and seldom descends to the ground (though in this, as in everything else, the Emperor is also thoroughly unpredictable). The most obsessed of all Emperor obsessives, Ian R.P. Heslop, a former colonial administrator who regarded butterfly collecting as an extension of big-game hunting, constructed a net with a thirty-foot bamboo pole, giving him a total reach of some forty feet, though how much control he could exert over such a length is debatable. More effective, perhaps, would have been the method suggested by H.G. Knaggs (he of the morally uplifting *Lepidopterist's Guide*) – shooting butterflies at rest with dust shot or a judicious squirt of water. However, these unsporting practices never caught on in England. Unlike many otherwise rare butterflies, Purple Emperors never, even in the best of times, appear in large numbers, and their populations are scattered. As a result, even the most obsessive hunter of these beauties could never amass huge numbers of specimens: Heslop himself had 'only' 185 to his name.

Uniquely, the Purple Emperor obsession continues into the present day, thanks in large part to the efforts of the great enthusiast Matthew Oates, who served as the National Trust's butterfly expert for nearly thirty years. He presides over a splendid illustrated blog, The Purple Empire, devoted to 'H.I.M.', an invaluable resource full of the latest news, sightings and developments, and written in a lively, readable style all too rare among today's lepidopterists. Oates also organises 'Purple Emperor breakfasts', events in which the most disgusting 'delicacies' – everything from horse manure and fox scat to rotting bananas and stinking shrimp pastes – are laid out in an attempt to lure H.I.M. down from the treetops. Part of the fascination of the species has always been the contrast between its imperial purple beauty and its depraved appetites (another favourite food is dog faeces). Even the earliest collectors used to employ dead cats as Emperor bait. The scientific explanation for the Purple Emperor's strange tastes is that the male in particular needs various mineral salts to renew his vigour after mating. There are plenty of other butterflies with similar habits, but, being smaller and more discreet, they tend not to get noticed: the Adonis Blue, for example, is very partial to animal droppings.

One of my regular butterfly haunts is known for its Purple Emperors (as well as White Admirals and Silver-Washed Fritillaries) and when I'm walking there in the season, passers-by often ask if I've seen 'the Emperor'. In fact it took me a long time before I did, but my first sighting was unforgettable – and, as it turned out, my best. It was one of those days when the Emperor-chasers were out, and, as I walked on one of the paths, I noticed to my left, some way away, a bunch of them training their binoculars on a particular oak. Deciding against going to investigate, I carried on along the path and, a minute later, looking to my right, I saw, at the start of a narrow side path, a magnificent male Emperor, treading the margin of a nearly dry puddle, seeming not to care that I was within an arm's length of him. At one point he even flew at me as if he might perch on me, but he changed his mind and returned to his muddy investigations. I was watching him for several enchanted minutes before he finally lost interest and flew off, disappearing from view with a few effortless beats of his powerful wings. It was an exciting encounter – my first Emperor! – but it was also undeniably strange. There is something about the burly, flashy Emperor that is quite unlike any other British butterfly, something exotic and almost unnerving. I am always looking out for Emperors, at the appropriate time and in the appropriate places, but if the year passes without a sighting (as it often does) I feel no great sense of loss, as I surely would if I saw no White Admirals or Silver-Washed Fritillaries…

At this point, I should perhaps confess my own particular obsession. It is one that has grown upon me in recent years (when I have had enough time to pursue it) and has become stronger with each season. Hairstreaks, inconspicuous little creatures with tiny 'tails' on their hindwings, seem to me to be, in their quiet way, among our most beautiful butterflies. They are also among the most fascinating ones to watch, and the hardest to find. The most straightforward of them is the first to emerge – the spring-flying Green Hairstreak (mentioned several times above). It is not easy to identify in flight, but once it has settled and folded its wings, that vivid green underwing makes it unmistakable – and, when settled, it can be surprisingly tame, happy to pose in the most obliging manner, but is a lively and pugnacious little butterfly, one that guards its territory ferociously, flying up from its favourite perch to chase away any intruders. Most of the visual interest of the Hairstreaks is in their underwings, which show the distinctive thin 'hairstreak' in various forms – in the case of the Green Hairstreak, a line of delicate silvery dashes. A still more delicate edging of reddish chestnut enhances the beauty of those perfect green underwings.

The Purple Hairstreak, a butterfly of high summer, is the commonest of our Hairstreaks – but that is not to say that it's easy to find. It spends most of its life in and around oak trees, preferring to stay high up in the treetops, taking to the air in occasional lively communal dances when the sun is out. Individuals do descend to lower ground, however, and, once you know what you're looking for, can often, in the right kind of territory at the right time of year, be found feeding on bramble flowers. The butterfly has a subtle purple sheen on its dark upperwings, and its underwings are pale grey with a strong silver 'hairstreak', delicately patterned margins and an orange eyespot just by the 'tail' – a fine example of understated beauty. This understatement makes the Purple Hairstreak an easy butterfly to miss, especially in flight, when it looks quite nondescript – but wherever there are oak trees, or even a single oak, there could be Purple Hairstreaks. Seek and you might very well find, even in suburban parks and gardens.

The same might almost be said of the White-Letter Hairstreak, but its tree is the less common elm (less common since Dutch elm disease) rather than the oak, and it is much less abundant and more elusive than its purple cousin. Like the Purple, it prefers to live up in the treetops, taking in the sun, drinking the 'honeydew' exuded by aphids feeding on the leaves, and occasionally taking to the air. The best way to see it, we are told, is to train binoculars on the crown of a likely elm on the sunny edge of a wood, but for myself I have only ever encountered the White-Letter on the ground or nectaring – each time a glorious surprise. The underwing – which is all you're likely to see of it, as it folds its wings tight on settling – is of pale brown, with a white streak ending in the jagged W or M shape that gives the butterfly its name, a scalloped edging of orange and black, and the telltale Hairstreak 'tail'. This beauty managed to elude even the voracious net of the young Nabokov, as he recalls in *Speak, Memory*:

'I remember one day when I warily brought my net close and closer to an uncommon hairstreak that had daintily settled on a sprig. I could clearly see the white 'W' on its chocolate-brown underside. Its wings were closed and the inferior ones were rubbing against each other in a curious circular motion, possibly producing some small, blithe crepitation pitched too high for a human ear to catch. I had long wanted that particular species and, when near enough, I struck. You have heard champion tennis players moan after muffing an easy shot ... But that day nobody (except my older self) could see me shake out a piece of twig from an otherwise empty net and stare at a hole in the tartalan.'

The White-Letter Hairstreak, having adapted its diet to various disease-resistant hybrid elms, is beginning to turn up in unexpected settings, including gardens, so in high summer I live in hope that I might spot one on my suburban wanderings. The same goes for the last of the Hairstreaks to emerge – the even more elusive Brown Hairstreak, which is the largest and perhaps the most beautiful of them all, its golden-brown underwings marked with a strong double streak of white and displaying a subtly modulated range of pale browns and orange, with a touch of black towards the 'tail'. Like the other tree-dwelling Hairstreaks, it is reluctant to leave the treetops, and if you are lucky enough to see one, it will most likely be a female that has come down on a sunny day to lay its eggs. Beautiful but easily overlooked, tantalisingly elusive, yet liable to turn up unexpectedly, these Hairstreaks are the stuff of dreams and obsession.

The Black Hairstreak lives almost exclusively in odd patches of woodland between Oxford and Peterborough, though it has recently turned up unexpectedly in Surrey. Perhaps it will be found elsewhere before long; like all the Hairstreaks, it is liable to pass unnoticed if no one is looking for it. I have visited one of the Black Hairstreak's haunts, without any luck. One day, when my obsession moves on to its next stage, I shall no doubt make a serious expedition in quest of it.

Flying through our dreams and on into our imaginative literature, music and art, and making a direct appeal to our emotions and our sense of being in the world, butterflies fascinate us, delight us, and can obsess us. But underlying all these emotional and imaginative engagements is one feeling that infuses all the others – the sense of enchantment. It is an enchantment of which the principal element is surely their sheer beauty...

On the Beauty of Butterflies

The beauty of butterflies is a major part of their appeal, even to the most narrowly science-minded lepidopterologist of today (or so I'd like to think). It is a beauty that comes in various forms, from the extravagantly spectacular beauty of tropical species to the quieter, understated beauty of British butterflies, which are generally smaller and less dramatically coloured and patterned than those of more southerly climes. There are two notable exceptions to this: the Swallowtail and the Purple Emperor. The English Swallowtail is certainly large – larger even than the Emperor – but it does conform to the general rule in being smaller and darker than its continental cousins. The Purple Emperor, on the other hand, is not only large but undeniably flashy, its purple-sheened upperwings and dramatically patterned underwings looking more tropical than English. It is always a thrill to see one of these scarce and highly elusive creatures, but it does feel rather more like watching an exotic in the butterfly house than a native species in its own habitat.

Both of these giants are strongly coloured and patterned, as are some much more common species, such as the Red Admiral, Small Tortoiseshell and Peacock (on their upperwings), but British butterflies with a strong single colour are exceptional: the sulphur-yellow Brimstone, the emerald underwings of the Green Hairstreak, the upperwings of some of the Blues. The particular beauty of British butterflies has, I would argue, more to do with pattern than with strong colour – and, to my eye, at least as much to do with the infinitely subtle patterning of the underwings as with the generally bolder effects of the upperwings. There is a good reason why the upperwings tend to be more dramatically marked than the underwings: they can serve to ward off predators, either by displaying warning coloration or by suddenly exposing a flash of something unexpected (an eye, or some illusory shape) to startling effect. Underwings are more often useful as camouflage and so tend to be more subtly marked, giving an impression of tree bark, lichen or dead leaf, or creating a visual effect that blends with the ground or the predominant vegetation. Some patterns, such as the silver wash (on delicate green) of the well-named Silver-Washed Fritillary, almost defy description.

A vast range of adjectives can be employed to evoke the patterns of coloration displayed by British butterflies – speckled, dappled, mottled, marbled, freckled, spangled, jewelled, chequered, grizzled – and pattern can take the form of specks, spots, studs, fine lines ('hairstreaks'), dots, 'eyes', ringlets, mosaic... A butterfly that beautifully demonstrates most of the possibilities of decorative patterning is the Painted Lady. The upperside has a ground colour that can range from dull orange to sand to something close to salmon pink. The forewings are tipped with black and white, and towards the body the ground colour is marbled with black. The slightly scalloped wing margins are marked with complex patterns of alternating black and white or black and ground colour. The hindwings are lightly marbled in brown and, towards the margin, marked with an inner row of dots, black with a pale orange corona around the lower dots, and an outer row of smaller, elongated dots forming part of the pattern of the hindwing margin.

And that is just the upperwing; the underwing is where the patterning really takes off. Here, the ground colour is a paler version of the upperwing, and the forewing partly echoes the patterning of its upperwing equivalent, but only up to a point: a kind of silvery-greeny-grey replaces the black of the wingtips, and two short black and white streaks, which have no analogue on the upperwing, run into the wing from the leading edge. The silvery-greeny-grey of the wingtips becomes the ground colour of the hindwing, which echoes the upperwing only in having a similarly positioned row of dots towards the wing margin – but these are 'eyes' of a slightly fuzzy blue and black, each one circled with pale yellow and a thin dark outer edge, and the one nearest the top of the wing 'blind'. Nearer the body, the ground colour becomes more grey than green, and the pattern is a delicately tinted, complex mosaic of small, variously rounded shapes between bands of white, with a small band of white and dark grey towards the top of the wing echoing something of the two bands that mark the forewing. It is a bravura display of subtle and intricate beauty (quite belying the name 'Painted Lady', though that fits the upperwing well enough). One of my earliest butterfly memories is of gazing in wonder at the underwings of Painted Ladies nectaring on Buddleia on that enchanted patch of ground in coastal Kent described in 'My Butterfly Life'. Amazingly, given all that coloration, the Painted Lady is remarkably effective at disappearing completely when it lands on bare ground. As if by magic.

'Glory be to God for dappled things,' wrote the Jesuit priest and highly individual poet Gerard Manley Hopkins in his poem 'Pied Beauty' (1877). It is a line that often occurs to me when looking at that common but uncommonly beautiful butterfly, the Speckled Wood. If ever there was a dappled thing – and one worth a *Gloria* – the Speckled Wood is it, though Hopkins seems not to have had butterflies in mind when he wrote his poem: a brindled cow, a stippled trout and 'finches' wings' are as specific as he gets. The Speckled Wood, known in earlier times as the Wood Argus, could fittingly have been called the 'Dappled Wood'. The creamy speckles on its chocolate-brown upperwings look exactly like dots of sunlight dappling the shade of a woodland ride. And the Speckled Wood is indeed a woodland butterfly – or it was originally, before it began its remarkable spread into more marginal habitats, including suburban edgelands and gardens, wherever there are sunlight and shade, grasses and brambles. There is quite a lot of variation in the markings of the Speckled Wood's wings, but in all cases that dappled effect, enhanced by a slight blurring of the edges of the speckles, is dominant. As well as the speckles, the upperwing has eyespots – a small, beady one at each wingtip and three larger ones, circled in cream, on the lower margin of the hindwing. The underwings are a subdued, subtly modulated version of the upperwings, with something like a bark or lichen pattern on the underwing and the 'eyes' much reduced. With its subtle, understated beauty, the Speckled Wood is in a sense the archetypal English butterfly; it is surely flying on Moses Harris's enchanted woodland ride.

Green is an uncommon colour among British butterflies, and the Green Hairstreak is the only instance of it as a strong single colour over an entire wing surface. The moss green of the Orange Tip's underwing is also unique, but something like it can be seen in the ground colour of the hind underwing of the Silver-Washed Fritillary, where it looks as if it has been laid on in thin watercolour washes, then streaked with silver-white and mottled with faint, but darker green, eyespots, to strikingly beautiful effect. A similar, still paler green is to be found, spangled with white, on the same wing of the slightly misnamed Dark Green Fritillary. And, on a much smaller scale, the Silver-Spotted Skipper, one of the last of our butterflies to emerge, has underwings of pale olive green dotted with silver. Other than that, despite the prevalence of green in our landscape, there is little of it to be found on our butterfly palette.

The single strong colour we are most likely to see is blue: the violet-tinged cerulean of the fairly abundant Common Blue and the paler, less intense tint of the increasingly common Holly Blue, not to mention the blue of the Peacock's 'eyes' and the edging of blue spots that heightens the beauty of the Tortoiseshell's wings. The colour is at its most intense on the upper-wings of the male Adonis Blue, a sadly scarce (but recovering) downland species. This shimmering metallic blue, sometimes turquoise, sometimes violet, is like nothing else to be seen among British butterflies, and it packs a powerful aesthetic, even emotional punch, at least among those of us who are especially drawn to blue. The Adonis's dazzling colour is, like much butterfly colour, the product of nanostructures in its scales, adapted to disperse all other wavelengths of light. It is an illusion, but this is true of most of the blue in nature: the blue of sky and sea, even the blue of eyes. Despite blue being a colour that is everywhere in nature, what we see is rarely the product of pigmentation: as painters knew well, before the coming of modern paint technology, true blue pigments were rare and costly, having to be created from ground-down lapis lazuli. It was reserved for the painting of the Virgin Mary's robes and the skies of Heaven.

Not all butterfly colours are produced by light reflected from specialised nanostructures; some of the more common ones are created from pigments, notably melanin. As well as black and brown, this, it is thought, is also the source of yellow and orange colours (and combinations can produce, for example, the green of the Orange Tip's underwings). Stepping back and taking an overview of the colours of our native butterflies, there is one colour that is clearly more abundant than any other – brown. This is the colour of the bodies of most of our butterflies, of the inner parts of the wings where they join the body, and of the soft, velvety 'fur' that in some species covers those parts. It is also the colour of earth and wood, both of which figure large in the English landscape, and is the basic colour of most of the group of our butterflies known as the 'Browns'. But 'brown' is an inadequate word for the range of tones displayed by butterflies' wings: from the gold tint of (some) Skippers and the dull orange ground colour of Fritillaries, through the various shades of sand and khaki, buff, fawn and grey-brown seen mostly in patterned underwings, to the flat woody brown of the Meadow Brown, the rich chocolate of the Brown Argus and the near-black of the Ringlet's upperwings and the Peacock's underwings. Consider the range of browns to be seen on one common butterfly alone – the Tortoiseshell, with its bright orange upperwings marked with dark brown near the body, dots of near-black and an edging of pale grey-brown, and its underwings displaying a range of browns from the palest buff to something close to black. Butterflies demonstrate, like no other living creatures, the beauty and chromatic range of brown, a colour that is too easily taken for granted as little more than a ground upon which other, less subtle colours may show off their lustre. It is their very brownness, in all its rich variety, that gives British butterflies their distinctive understated character and quiet aesthetic appeal, something very different from the wider and more vivid colour palette displayed by the butterflies of sunnier climes.

So far I have considered the beauty of butterflies purely in terms of colour and pattern, as if they were so many textile designs. It is no coincidence, then, that many of the early Aurelians were involved in the silk industry; it's easy to see how the patterns and colours, the lustre and sheen of a butterfly's wings would appeal to a textile designer's eye. However, butterflies are of course living things, moving freely through the air, and their movements are very much a part of their charm and their beauty. When we see a butterfly in flight we are seeing all that surface beauty of colour and pattern enhanced, modified, concealed and revealed, set in motion by the butterfly's movements. Butterfly beauty, even when the butterfly is 'at rest', is kinetic (which is one of the reasons why specimens pinned on a board are so sadly inadequate).

The way butterflies fly varies greatly from species to species: from the Purple Emperor soaring, swooping and charging about the treetops, to the Mountain Ringlet rarely venturing more than a foot off the ground and landing every time anything brown catches its eye. The kind of graceful, swooping flight exemplified by the Purple Emperor is, in less overtly aggressive species such as the Silver-Washed Fritillary, Comma and White Admiral, very beautiful to watch. Other butterflies have a busy, darting flight – none more so than the Skippers – while some larger species fly fast and purposefully: Dark Green Fritillaries hurtling over downland, Painted Ladies and Clouded Yellows on their migration flights, even Red Admirals can put on an impressive turn of speed. Towards the other end of the spectrum are the ineffective fluttering of the Wood White and the weak, floppy flight of the Large White and Meadow Brown. But the most characteristic and perhaps the most cheering feature of many butterflies is their dancing flight, rising and falling over grassland and along hedgerows, circle-dancing around trees and shrubs. Grassland is the best place to enjoy it. It is a gloriously heart-lifting experience to lie as close to the ground as possible in a sunny flower meadow or on well-cropped downland and watch the busy, endlessly moving insect world at work all around: bees buzzing from flower to flower, flies of all kinds going about their various businesses, beetles, ants, bugs and spiders roaming the ground, and amid all this the butterflies, some taking nectar from their favoured flowers, some following erratic chemical trails undetectable to us, and many rising and falling above the grass and flowers in endlessly moving, constantly changing patterns of dancing flight. There are chases too, and courtship and sparring (often hard to tell apart), and sometimes a butterfly, or a pair, shadow-dancing in a double helix, rising up and up into the blue. It is the dance of nature, the dance of life. In *Specimen Days* (1882), the American poet Walt Whitman gives a rhapsodic account of such a scene, a 'butterfly good-time', in a hay field:

'Over all flutter myriads of light-yellow butterflies, mostly skimming along the surface, dipping and oscillating, giving a curious animation to the scene. The beautiful, spiritual insects! straw-colour'd Psyches! Occasionally one of them leaves his mates, and mounts, perhaps spirally, perhaps in a straight line in the air, fluttering up, up, till literally out of sight. In the lane as I came along just now I noticed one spot, ten feet square or so, where more than a hundred had collected, holding a revel, a gyration-dance, or butterfly good-time, winding and circling, down and across, but always keeping within the limits. The little creatures have come out all of a sudden the last few days, and are now very plentiful. As I sit outdoors, or walk, I hardly look around without somewhere seeing two (always two) fluttering through the air in amorous dalliance ... As I look over the field, these yellow-wings everywhere mildly sparkling, many snowy blossoms of the wild carrot gracefully bending on their tall and taper stems...' (George Bristow, the 'father of American classical music', evoked a similar scene in music, in *The Butterfly's Frolic*, the scherzo of his third symphony.)

From Whitman's New England to old England, the scene is much the same wherever butterflies get together in sociable abundance.

To return to ground level in our flower meadow or patch of downland, mixed in with the complex, ever changing web of insect motion is a soundscape of chirring grasshoppers, buzzing bees and singing birds (perhaps a lark overhead, a cuckoo, a whitethroat) and the smells of the warm dry earth, of flowers and herbs and grasses. We are experiencing the life of a little ecosystem, a small world in balance, going about its business oblivious of us, but beautiful to our senses. The essence of that little world is movement, most perfectly expressed in the dancing of the butterflies.

Only one British butterfly has 'Beauty' in its name – the Camberwell Beauty. It has had various other names – the Willow Butterfly (as in Benjamin Wilkes's *English Moths and Butterflies*), the Grand Surprize (a marvellous name), the White Petticoat, and of course the Mourning Cloak, which is still its American name – but 'Beauty', in the end, it had to be. And 'Camberwell Beauty' because the first two specimens known to English aurelians were taken in the summer of 1748 in Cool Arbour Lane, near Camberwell, then a village in open countryside. Cool Arbour Lane is now a traffic-choked highway, Coldharbour Lane, and Camberwell's fields and woods long ago disappeared under street after street of housing – but the name of its famous butterfly lives on. Even in the eighteenth century the Camberwell Beauty was 'one of the scarcest Flies of any known in England', and so it remains, though there are occasional years when hundreds of Camberwell Beauties turn up. Even so, seeing one is always a 'grand surprise'. It certainly was to the poet and memoirist Siegfried Sassoon, as he related in *The Old Century and Seven More Years* (1938). He was sitting in the attic of his Kentish home one summer day when he became aware of the flutterings of a butterfly trapped behind the gauze covering the skylight...

'By standing on a chair – which I placed on a table – I could just get my hand between the gauze and the glass. The butterfly was ungratefully elusive, and more than once the chair almost toppled over. Successful at last, I climbed down, and was about to put the butterfly out of the window when I observed between my fingers that it wasn't the Small Tortoiseshell or Cabbage White that I had assumed it to be. Its dark wings had yellowish borders with blue spots on them. It was more than seven years since I had entomologically squeezed the thorax of a "specimen". Doing so now, I discovered that one of the loftiest ambitions of my childhood had been belatedly realised. I had caught a Camberwell Beauty.'

By way of contrast, Kingsley Amis, in the first stanza of his poem 'To H.' (1991), recalls how, at the age of ten, in his grandmother's Camberwell garden, he saw a Camberwell Beauty and thought 'What else would you expect? Everyone knows Camberwell Beauties come from Camberwell; That's why they call them that.' Only later did he discover how astonishingly lucky he had been.

As a boy, I used to dream of seeing a Camberwell Beauty (from time to time I still do), but I had to wait until many years later, when, on a visit to Canada, I had my own 'grand surprise'. It was in Montreal – indeed on Mount Royal (Mont Réal), the great hill that dominates the city and is laid out as a magnificent 'natural park', the work of Frederick William Olmsted, the creator of New York's Central Park. There, suddenly, on a dry earth path, with its wings fully spread, was the butterfly of my dreams – the very last thing I had expected to see on a cool April day with the ground still patched with unthawed snow. The beautiful creature was understandably torpid, and very nearly – wonder of wonders – walked onto my outstretched finger: four feet were on before it changed its mind and, summoning its energy, flew off. Later in my Canadian travels I saw several more 'Mourning Cloaks', but they were all in strong flight, so I could not properly enjoy them. Little, if anything, in my butterfly life before or since has topped the thrill of that first encounter with a living Camberwell Beauty, that grandest of surprises.

It is no wonder this magnificent butterfly was named a 'beauty': its upperwings of dark chocolate brown, studded with blue and edged with creamy-white (or yellow), are spectacularly gorgeous. The best description of it is in a poem that Vladimir Nabokov wrote when he was a student at Cambridge. 'Butterfly (*Vanessa antiopa*)' describes a butterfly that is 'velvety-black, with a warm tint of ripe plum' – exactly the colour and texture of the Camberwell Beauty's upperwings. This 'live velvet' is bordered with 'a row of cornflower-azure grains, along a circular fringe, yellow as the rippling rye'. The butterfly 'has perched on a trunk, and its jagged tender wings breathe, now pressing themselves to bark, now turning toward the rays' – a precise description of a butterfly's movements as it warms itself. Then it's back to the beauty of those upperwings, which Nabokov encapsulates in one brilliant image – 'a blue-eyed night is framed by two pale yellow dawns'. Perfect.

There are beauties too among our smaller and commoner species, though none has 'Beauty' in its name. Consider the Orange Tip: this small butterfly – white with unmissably bright orange tips to its wings – is one of the most cheering sights of spring. It is only the male who sports the orange wingtips, but males are the ones you are most likely to see, especially early in the season, when they are always on patrol, urgently seeking out females (whose wingtips are edged with charcoal grey: no orange). A close look at the male Orange Tip's upperwings reveals the subtle shading and patterning which give that white-and-orange coloration its particular beauty. The bright tips of the forewings are edged with a dark blackish-brown margin and dotted on the inner edge with a speck of the same shade. The outermost wing edges are delicately patterned with the same colour alternating with white, a pattern echoed in a fainter, sparser edging of the hindwings. Here the ground colour is modified by a faint greyish marbling derived from what lies under – the exquisite green mottling of the hind underwing. This patterning seems to resemble lichen, but it functions as highly effective camouflage for the butterfly when it settles on its larval food plants, Jack by the hedge and cuckoo flower, or on cow parsley or any umbellifer. Folding its wings vertically and raising those delicately marked green-and-white hindwings over its forewings, the orange tip has no trouble hiding in plain sight – a particularly useful ability for the female, which spends far less of its time flying than does the ever-amorous male.

It would be possible to fill many more pages with descriptions of butterfly beauty: every species, even the most superficially dull looking, has something beautiful about it when looked at with due attention. But now it is time to look beyond appearances, at the ways in which we give butterflies meaning, and the lessons which we might learn from observing their butterfly ways.

Meanings and Lessons

Perception is a creative act. Being inescapably human, we cannot help but project human meaning and significance onto what we see, to interpret it into something we can assimilate. We are not passive receivers of data but, rather, transformers, actively making sense – or sometimes nonsense – of what we are perceiving. We seek out shape, pattern, meaning; we want to fit whatever is there into our human world; what other world can we truly know? This leaves all nature wide open to our interpretation and our humanising, assimilating impulse. Anthropomorphism (the attribution of human behaviour and motives to the non-human realm) is frowned upon, especially among naturalists, but, in a broad sense, it is what we cannot help but do, whether or not science approves, and whether or not we are conscious of doing it. We anthropomorphise everything: we could not understand it (comprehend it) otherwise. Perception is interpretation.

A basic form of interpretation is identification – especially important to naturalists, but not only to them. In the human sphere, identification is an essential condition for paying due attention. It is what brings our fellow humans alive to us, makes them something more than 'just another' person. There is a famous passage in Proust, where the narrator, Marcel, anxious about his grandmother, races to Paris to see her. When he arrives, she is not expecting him and he witnesses, as it were, his own absence. In that absence, what he sees, shockingly, is not his grandmother but a florid-faced, mad old woman, sitting in a chair reading. He has seen her but, for that moment, he has not identified her, and so has not paid her that loving attention which will transform her instantly from a mad old woman into his beloved grandmother. Attention – ideally, loving attention – is what makes us what we are, and what brings our human world alive with meaning. It is indeed, in Iris Murdoch's phrase, 'the characteristic and proper mark of the moral agent'.

Similarly, albeit in a much less dramatically charged manner, we give life and meaning to what we are seeing when we identify a butterfly, distinguishing it from the generic and recognising its unique identity. This is one way in which we fit the living creature into our human world – in this case into our scientific system of classification of species. But we have found many other ways, thanks to our urge to project human meaning onto nature. One, rather surprisingly, is in the naming of species. Back in the days when every gentleman (and many ladies) had a thorough grounding in the classics, the Latinate naming of butterflies was at once a scholarly and a poetical affair. Drawing on a wide knowledge of classical mythology and literature, and permitting a range of allusion that would be apparent to the educated (if not to anyone else), it added new layers of meaning to the names of our butterflies. Like the illustration of butterflies in the Aurelian age, their naming also showed how porous the border between science and the arts was in the early years of lepidoptery.

Those classically minded taxonomists of the eighteenth century used their learning to describe butterflies in terms of appearance and behaviour, or to draw analogies, often obscure, with figures from classical mythology. The larger, more obviously beautiful butterflies were usually given female names, and a family of them was named Nymphalidae after the nymphs of classical mythology; the less showy 'Browns', on the other hand, were seen as masculine, and a family of them was named Satyridae after the wood-dwelling satyrs. The Three Graces inspired the naming of three Fritillaries (the Dark Green, Pearl-Bordered and Heath, named after Aglaia, Euphrosyne and Athalia), and the White Admiral, *Ladoga camilla*, was named after Camilla, a warrior princess in the Aeneid. The Painted Lady was named *Cynthia*, an epithet of the goddess Artemis and the name of a popular all-purpose muse in English poetry. Sometimes the classicists got carried away with themselves, as when naming that cheeriest of butterflies, the Gatekeeper, *tithonus*, after the tragic figure who begged the gods for eternal life and found himself cursed with endless old age and an inability to die (as beautifully described in Tennyson's poem *Tithonus*). And does the Meadow Brown, that ubiquitous grassland butterfly, really give the impression of a small ghost, which is what its Latin name, *Maniola*, suggests? Similarly, does the inconspicuous little Dingy Skipper deserve to be named *Erynnis*, after the Furies, the Erinyes?

The origins of the Latin name for the Red Admiral, *Vanessa atalanta*, are surprising. Vanessa is a name that sounds as if it has been around for ever – the kind of name that might be found in Latin love poetry – but it was in fact invented by the poet and satirist Jonathan Swift as a pet name for Esther Vanhomrigh, the unfortunate young woman, twenty-two years his junior, who was the object of his obsessive love for seventeen years, until he moved on to another Esther (Esther Johnson, nicknamed Stella). Swift arrived at 'Vanessa' by conflating the 'Van' of Vanhomrigh with 'Esse', a pet form of Esther. The name caught on, and it was not long after its first appearance in print, in Swift's poem 'Cadenus and Vanessa', that the Danish naturalist Fabricius named the genus, to which the Red Admiral belongs, *Vanessa* – a properly Latin-sounding name, despite its origins. Linnaeus subsequently gave the Red Admiral its grand binomial, *Vanessa atalanta*. Atalanta was a figure from Greek mythology, a formidable virgin huntress and athlete who was notably reluctant to marry, and her name was, as it were, in the air in the eighteenth century: Handel's pastoral opera *Atalanta* was first performed in 1736, and the score sold more copies than any of his others in his lifetime (it includes the beautiful aria 'Care selve' – 'Dear woods'). Before that, Atalanta's athletic and hunting exploits had made her a popular subject for artists, most notably Rubens, who painted her several times. Atalanta's name even appears, once, along with Vanessa's in 'Cadenus and Vanessa' – 'When lo! Vanessa in her bloom/Advanced, like Atalanta's star.' This must be the only instance in English poetry of the two halves of a Linnaean binomial turning up in consecutive lines…

A few butterflies – the Adonis Blue, the various species of Argus (after the many-eyed shepherd of Greek myth) – have a classical element in their common English name, but for the most part these names are home-grown. Though not generally quite as beautiful as our names for wild flowers, traditional English butterfly names have a poetry and charm to them, and are often wonderfully descriptive: 'silver-washed' for the underwings of the Silver-Washed Fritillary is perfect, as are 'marbled' for the Marbled White, 'silver-studded' and 'silver-spotted' for the Blue and Skipper so described, and 'pearl-bordered' for that Fritillary. The word 'fritillary' – also a flower's name, of course – is in itself both beautiful and descriptive, ultimately deriving from the Latin for a patterned dice box.

'Hairstreak', too, is the perfect word for the fine lines on those butterflies' underwings, and 'comma' for the curved white mark on a Comma's dark underwing. Other butterfly names exalt their species – none more so than the Purple Emperor, the Queen of Spain Fritillary and the Camberwell Beauty. Such poetry is absent from more recent naming, which is a purely scientific affair, prosaically descriptive, often including the name of the first identifier of the species, as in Berger's Clouded Yellow and Réal's Wood White (or indeed Nabokov's Satyr). In the eighteenth-century golden age of butterfly naming, the field was wide open for the Aurelians and the Linnaeans to let loose their imaginations, and their classical learning, and the results were often rather wonderful.

Our irresistible urge to interpret the natural world in human terms has given meaning to butterflies in general, but has also invested individual species with particular meanings – none more so, perhaps, than the Red Admiral. This is the butterfly that turns up most often and most recognisably in the works of Nabokov. In *Pale Fire*, an extraordinary work made up of a narrative poem by a fictional American academic, John Shade, and a supposed commentary on that poem by a colleague who is at least half mad, the Red Admiral appears in two guises. In Shade's poem, the butterfly is associated with his adored wife – 'My blest/My Admirable butterfly!'– and with his impending death. At the end of the poem, Shade looks for his wife just before catching sight of the Red Admiral that will be among the last things he sees: 'A dark Vanessa', wheeling in the sun before settling on the sandy ground.

For all its beauty and its lovely Latin name, the Red Admiral has long been associated with death and the dark side. Its most conspicuous colour combination of red and black has something diabolical about it: the red of hellfire and the Devil himself against the darkness of the pit. The sudden appearance of a Red Admiral might easily be taken as an ill omen – and it is a bold butterfly with a strong, dramatic presence, a creature that can easily be taken for a 'messenger'. In Russia the Red Admiral got a reputation as the 'butterfly of death' after vast numbers appeared in 1881, the year of the assassination of Tsar Alexander II. There are markings low down on the dark underwings that can be read as a distorted '8' and '1' (the '8' nearer to the body), so that the date 1881 might be read, with a little imaginative effort and wishful thinking, across the butterfly's spread wings – an example of the book of Nature literally imprinted with human numbers.

Nabokov characteristically refers to this butterfly as the Red Admirable rather than the Red Admiral. He seems to have believed – as many still do – that 'Admirable' was the original form, later corrupted into 'Admiral'. In fact, the reverse is probably closer to the truth: 'Admiral' was the earliest recorded name, 'Admirable' a later version that gained little currency except with those who, reasonably enough, liked the sound of it. 'Admiral' was the common name for this butterfly across northern Europe, for no very clear reason. The early aurelian James Petiver gave the name 'Admiral' to several other butterflies as well as the Red, but it seemed to suit the Red particularly well, as the band of red on the wings could suggest the scarlet sash of an Admiral in the Royal Navy (one of whose squadrons was called the Red). However, the resemblance is far from obvious, and the prevalence of the name 'Admiral' in other countries rather undermines the idea of a British naval connection. The word 'admiral' has an Arabic root, the word 'admirable' a Latin one, but perhaps there was a period when the two were interchangeable and both carried the meaning 'admirable'. That would certainly be the simplest explanation.

Butterflies seem to have accrued symbolic meanings very early (though these are often lost on us), and those meanings have continued to proliferate into our own times, long after Psyche took wing and Christianity dismissed the frivolous butterfly. In our own time, butterflies have come to represent all good things – freedom, beauty, peace, love, innocence, happiness, eternal life, you name it. Not for nothing is the butterfly described as 'the world's most popular insect'.

One of the most spectacular demonstrations of the symbolic power of the butterfly took place in Hyde Park, London, in the summer of 1969. At a free Rolling Stones concert (yes, free – those were the days), Mick Jagger marked the recent death of original band member Brian Jones by stepping forward and declaiming Shelley's 'Adonais: An Elegy on the Death of John Keats' – while a veritable swarm of white butterflies (*Pieris brassica*, the Large White) was released from several large boxes. Originally there had been 2,500 of them, but by the time of the release most had died from lack of oxygen, and only a few hundred remained alive long enough to take to the air. Most of those soon fell to the ground and expired. 'It was like the Somme,' recalled drummer Charlie Watts, who took a dim view of such flamboyant gestures. 'Like the Somme' was probably just rock-star hyperbole, but it was in fact a surprisingly apposite image, for the battlefields of the Western Front often witnessed prodigious summer snowstorms of white butterflies. The war artist William Orpen describes one at the Somme in August 1917: 'The dreary, dismal mud was baked white – and pure dazzling white. Red poppies and a blue flower [cornflower?], great masses of them, stretched for miles and miles. The sky was dark blue, and the whole air up to a height of forty feet thick with white butterflies. Your clothes were covered with butterflies, it was like an enchanted land...'

Another visitor wrote that the great profusion of butterflies 'was as if the souls of the dead soldiers had come back to haunt the spot where so many fell. It was eerie to see them. And the silence! It was so still that I could almost hear the beat of the butterflies' wings.' In one of the most famous novels of the First World War, Erich Maria Remarque's *All Quiet on the Western Front*, the narrator, Paul, who collected butterflies in his boyhood, notices their presence amid the devastation of the battlefield: 'The grasses sway their tall spears; the white butterflies flutter around and float on the warm wind of the late summer.' Later in the novel, he sees 'brimstone-butterflies, with red spots on their yellow wings. What can they be looking for here? There is not a plant or a flower for miles. They settle on the teeth of a skull.' A grisly image of the endurance of nature and beauty amid all the man-made horror.

Clearly the unfortunate butterflies released in Hyde Park were intended to carry symbolic weight (and hang the consequences for the mortal butterfly), and that meaning was very close to the Greek association of the butterfly with the human soul liberated from the shell of the body. The gesture would have been considerably more eloquent if only one butterfly had been released to fly away to freedom, but rock and roll does not lend itself to understatement. In recent years, however, the release of a single butterfly, or sometimes more, has become quite a popular feature of funerals, and the intended symbolic meaning here is clear enough. It is equally clear that when butterflies are released at weddings – as is also becoming popular – they signify love and happiness. The species most commonly released at both weddings and funerals (and specially bred for the purpose) are the Painted Lady and the spectacular, exotic Monarch – never the altogether too common Large White. I once had the unsettling experience of coming across a Monarch happily nectaring on ragwort at the edge of a park near my home. This is the kind of thing to set a butterfly lover's heart racing, but at the same time my brain was telling me that this was far too fresh a specimen to be one of the few Monarchs that do, from time to time, turn up in Britain, blown over from North America. In fact, as I later discovered, there had been at least one butterfly release at a wedding that weekend, and this fine specimen must have flown my way from somebody's nuptials.

Such is their symbolic power and universal appeal that butterflies are now everywhere, a self-evident

Good Thing, providing a quick hit of low-impact spiritual uplift. The artist Damien Hirst, unsurprisingly, has made use of them on an industrial scale in his work, producing a series of pieces made with geometrically arranged butterfly wings – just the sort of thing the Victorians loved – and filling two windowless rooms full of living butterflies. The Tate Modern, where the exhibition was held, has since revealed that some nine-thousand of them died, many trodden underfoot by visitors. Butterflies have not been well served by conceptual 'artists' and rock stars with big ideas.

In the wider world, beyond Damien Hirst's cynical exercises in kitsch, butterflies are a universally popular decorative motif on greetings cards and wrapping paper, in textiles and jewellery, art prints, T-shirts and tattoos, packaging and advertising, everywhere. Whatever it is, put a butterfly on it. In recent decades butterfly houses, in which exotic specimens fly freely, have become increasingly popular, as has butterfly gardening, planting nectar-rich flowers (and leaving 'wild' patches) with a view to attracting butterflies. At the same time, butterfly rearing kits have become widely available: these enable children to watch caterpillars grow and turn into chrysalids, and then emerge as butterflies, usually Painted Ladies. This might not do much for conservation, but it must surely have given some children a vivid experience of the wonder of metamorphosis. Whatever the moral misgivings of the Church, the butterfly has become in the course of time a symbol of all good things.

So, we have projected much human meaning onto butterflies – but do we also get something human back from them? Do they, as it were, have something to teach us? I have explored what we might learn from watching them in terms of gaining perspective, focusing concentrated attention, standing still, and taking informed delight in what is around us, but what else can we learn from these beautiful and fascinating creatures? Could it be that simply watching them as they live their lives in the wild might give us some hint about how to live our own lives? It is a big claim, but I think the characteristic behaviour of butterflies does point us towards a possible way of living, of *living lightly*.

It is no doubt true that all non-human animals live in the moment – something we humans find extremely difficult – but it seems somehow especially true of butterflies: their flying season is so brief and their activity so dependent on the vagaries of weather, their bodies and wings are so fragile, and their short-lived beauty seems to make their little lives so much more precious. Why should a butterfly not seize the day and enjoy, in the most literal sense, its moment in the sun? Why should we, heavy-footed and long-lived creatures though we are, not try to do something similar? There is plenty to hold us back, from our everyday cares and distractions to our awareness of time as having a past and a future as well as a present, but the very act of watching butterflies can itself offer us an experience of living in the moment, as described so eloquently by Nabokov. We cannot live like butterflies, of course, but their behaviour might give us some hints as to how we could live our life more lightly, and more in the moment.

Like the bee, the butterfly typically flies from flower to flower, never staying long in one place but taking what it needs and moving on. This behaviour earns the bee the reputation of industriously 'improving each shining hour', but the butterfly, with its less purposeful-seeming flight, is seen as frivolous and time-wasting. Besides, it is only feeding itself, not obligingly making honey for us humans to enjoy. The aimless fluttering of the butterfly from flower to flower is what we have in mind when we talk of someone having a 'butterfly mind' – a mind that has been formed (or prevented from forming) by flitting from one thing to another, never settling for long, never probing deeply into one thing. Such a person, it is thought, might end up with wide but superficial knowledge and experience, rather than the deep insight gained by more concentrated and prolonged attention. To this I would reply, what makes 'wide but superficial' of less value than 'concentrated and prolonged'? Isn't the 'butterfly mind' much like the 'well-furnished mind', something that used to be highly valued? To have a broad perspective, to be able to make connections and pick up on a wide range of allusions and references, surely makes for a richer mental life than to know an awful lot about one subject area only – just as to be widely read or to have heard a lot of music is surely preferable to reading only a handful of authors and listening to only one kind of music. The more areas of life we have touched on, if only superficially, the more of what life has to offer we are able to enjoy. The terms 'dilettante' and 'amateur' are used pejoratively, but what is so bad about delighting in things (the dilettante) or loving them (the amateur)? We are back to those 'three good things': perspective, attention, ***delight***. The sniffy, suspicious attitude to the 'butterfly-minded' derives from a puritanical disapproval of pleasure for its own sake, of lack of seriousness. It has its roots in the traditional Christian disapproval of butterflies themselves as frivolous, useless creatures. By extension, those people who are reluctant to settle to one thing and bore (*le mot juste*) deeply into it must be regarded as lacking in seriousness and altogether too keen on extracting pleasure and interest from a wider range of what life has to offer.

Happily, there have always been some who have seen the attraction and value of the butterfly mind. In reaction to the relentless moral earnestness and machine-driven busyness of late Victorian England, such writers as Oscar Wilde and Max Beerbohm, and painters like Whistler (whose signature was a stylised butterfly), represented a refreshing refusal to take life (and ideas) too seriously, while in France the figure of the *flâneur* sauntered into the picture. The *flâneur*, strolling aimlessly but observantly about the urban streets and boulevards, 'botanising the asphalt' (in Walter Benjamin's phrase), offers a fine image of the butterfly mind at large, living lightly. They might be seen as pioneering psychogeographers, but the *flâneurs* were also like the best kind of naturalists translated to an urban setting, observing and enjoying whatever in the whole environment, the urban *terroir*, caught their attention or piqued their interest, and moving on – at a stroll, of course.

The butterfly mind, being flexible, receptive and resilient, is, I would argue, peculiarly well suited to our times. The mind that can travel lightly across the vast terrain opened up by the internet, skimming the surface and taking from it only what it wants or needs, navigating between its points of light and avoiding the darkness between, never disappearing down one of the many delusive rabbit holes that litter the online world – surely such a mind is a decided advantage in a world that is increasingly hard to navigate. Such a mind is also better adapted to a working world where, increasingly, careers are built not by staying in one place and climbing the ladder but by moving from one thing to another, never staying too long before moving on to something new, sometimes seizing the opportunity of a complete career change; to the nimble butterfly mind this is a natural way of doing things.

The butterfly mind, by its very nature, is in no danger of becoming obsessively focused on one thing, one idea or one opinion; it is a mind that is happy to entertain an idea, but not to invite it to move in and stay indefinitely. Like the Stoic philosophers of the ancient world, the butterfly-minded are less interested in being right or wrong, in winning or losing an argument, than in attaining and preserving the lucid state of robust equanimity, of balance, that the Greeks called *ataraxia*, a state that (ideally) enables us to live in the moment, to enjoy what can be enjoyed and to endure what must be endured. The butterfly mind is alert, like the butterfly itself, to threat, but also to possibility, to opportunity, to what is on the periphery as well as to what is at the centre of its attention. 'Clear vision goes with the quick foot,' writes Robert Louis Stevenson in one of his *Essays of Travel* (1905): we see most clearly when we are in motion, passing through; if we linger, that clarity will dissipate and lose its sharp precision. The light-footed, passing through, might be getting only an impression, but that impression might have at least as much truth in it as the more ponderous and systematically minded will discover in their methodical investigations.

If we look and listen, everything – nature itself – tells us to tread lightly on this Earth, to leave as small a footprint as we can, not to blunder about, careless of what is around us, but to pay attention to our environment, to appreciate its endless complexity and its vulnerability, to stop and look and wonder at how the creatures within it live their interwoven lives. The dance of the butterflies, in its graceful delicacy, lightness and tact, shows us a way of doing this, of living lightly.

Part IV
Into Butterfly Country

Where Are the Butterflies?

So, here we are in the mindful present. Where are the butterflies? It's a question even the most sanguine among us find ourselves asking with a slight edge of anxiety, as there is no denying that, in broad terms, butterfly numbers are tending to decline, if not from season to season, at least from decade to decade. In a bad year, one of those where spring is dull and summer washed out, the lack of butterflies is easily explained, but in a good year, when the sun is shining and all seems set fair, the sparseness is more disturbing. What is going on? Is the much-touted 'insect apocalypse' under way? This scenario, which came into play after a big research project in Germany found that insect populations in nature reserves there had declined by more than three quarters in three decades, can seem all too believable.

Though those findings have been contested, the evidence of our own experience suggests they are, at the least, along the right lines. Many of us are old enough to remember when car windscreens had to be wiped clear of splattered insects after a summer drive, while at night the headlights would reveal a swirling snowstorm of moths. That rarely happens now, except perhaps in the deepest, most untouched corners of the country. If insect numbers are indeed declining, there is no reason to expect butterfly numbers to be bucking the trend; many of our British butterflies are already at or near the northern extreme of their range, and several are hanging on only in isolated pockets of suitable habitat, either natural or man-made.

However, the overall pattern of decline is not the whole story, at least when it comes to butterflies. As ever, things are more complex than that: some species have increased in numbers over the past few decades; many conservation efforts have been successful; and some habitats, particularly suburban and semi-urban, have actually become more rewarding to the butterfly watcher. Butterflies are still there, even in a bad year, but we need to know where and when to look. And we need to broaden our idea of what 'butterfly country' is: it consists not only of pockets of especially butterfly-friendly countryside, often set aside as SSSIs (sites of special scientific interest) or nature reserves with strict conservation regimes. There is butterfly country to be found in far less obvious, and more easily overlooked, places. What follows is a guide, mostly practical, for the would-be butterfly watcher setting out in a spirit of curiosity to see what they can see...

Practicalities

The butterfly watcher of today is unencumbered by that infallible sign of eccentricity – the net – and need not even carry a knapsack. Butterfly watching is one of those pursuits for which, essentially, ***no kit at all*** is required. Of course, the same could be said of walking, and yet for many this activity seems to demand specialised clothing and all manner of equipment. Happily, we butterfly watchers will, in the nature of things, be pursuing our calling in generally clement weather and in less challenging parts of the landscape, so we have no need of elaborate protective clothing or high-performance boots. For myself, I prefer to give the impression of a *flâneur* who has strolled into the countryside by mistake, but likes what he sees, and intends to pay close attention to it. There is no reason why a butterfly watcher shouldn't dress in a civilised manner, as those early aurelians did.

So, all you need in the way of kit is:

1. A good pocket guide for identifying what you spot. The best is Jeremy Thomas's superb *Butterflies of Britain and Ireland*, which has been published in various editions over the years. You need one of the pocket versions, not the magisterial full-sized edition, with Richard Lewington's illustrations – the most beautiful and authoritative butterfly book of our time, but one that would require a pocket of eighteenth-century dimensions to hold it.

Lewington's own *Pocket Guide to the Butterflies of Great Britain and Ireland* is the most attractive of all compact guides. Do not, whatever you do, try to use a guide to the butterflies of Britain *and Europe* in the field: you will discover that British butterflies are but a drop in the vast, perplexing ocean of continental European species (they outnumber ours by eight to one). As you get more experienced, you will not need to carry a guide, but might find a small notebook useful for jotting down anything worth checking in the books when you get home. If you're fortunate enough to live in a county that has a reasonably up-to-date field guide to its butterflies, it is well worth seeking out, as it will give you much more localised information (about best sites, flying times, etc.) than any national guide (for US-specific guidebooks, please see the Resource List at the end of this book).

2. Binoculars. Though I don't make much use of them myself, they can certainly be good for seeking out those species that hang around in treetops, or for closing in on a specimen that has settled on a frustratingly distant perch and will not fly any nearer. If you'd rather not be seen with them, you can always stow them in your bag.

3. A camera, if you must. Most butterfly fanciers these days can be identified not by the telltale net of yesteryear but by the quantities of expensive optical equipment dangling from their necks (and the ubiquitous sun hat). If photography is your thing, photographing butterflies certainly offers some stimulating challenges, and can add extra interest to the watching. Also, if you reach the stage when you are getting serious and reporting sightings, it always helps to have a photograph as verification.

That's it! Butterfly watching bucks all contemporary trends (a recommendation in itself) by being the ultimate minimal-kit pursuit.

Another attractive thing about butterfly watching is that it is relatively easy to get fairly good at it. Not only is little kit required, and no physical prowess, but also there is, relatively speaking, little to learn. Consider this: there are only around sixty species of British butterflies (and a good number of them you are not going to see without making a special journey). Compare that with Britain's six-hundred-plus species of birds, one-thousand-six-hundred-odd species of wild flowers, twenty-seven thousand insect species – or indeed our two thousand five hundred species of moth. It does not take much time or effort to learn to recognise all the butterfly species you are likely to see. A little preliminary reading (and/or online research on, for example, Butterfly Conservation's excellent website) is a good idea, but not essential. In fact, even being able to identify species is not absolutely essential; you can experience at least the aesthetic pleasure and benefits of butterfly watching without necessarily having that knowledge. However, I am convinced that having at least some identification skills – knowing what you are looking at, what its habits are, and what distinguishes it from other butterflies – makes for an altogether richer and more rewarding experience. You cannot see far into the fascinating world of butterflies without some basic knowledge of different species and their ways. But don't worry – not only are there relatively few British species, there are not many of them that are seriously difficult to tell apart.

Butterfly watching is so easy that, if you're reading this on a fairly warm day with a bit of sunshine, you could start right now by stepping outside, into your garden (if you have one), or perhaps the nearest park, open space or patch of countryside, and looking around. The chances are that there will be a few butterflies about, even if they're only 'Cabbage Whites' (a generic term loosely applied to three species, but we won't go into that now). Observe their behaviour: how they fly, how often they settle and where, what they are doing – feeding on flower heads, basking in sunlight with open wings or perching with wings folded, laying eggs perhaps (if they're on a cabbage patch this is quite likely), interacting with one another, taking little flights together, mating even, if it's warm enough. Look out for one that has settled somewhere where you can get a good sight of it, slowly approach as close as you can (taking care, as ever, not to cast a shadow) and examine the markings on the wings. You will discover that there's a great deal more to those Whites than whiteness: if they were prized rarities, rather than commonplace sights that we take for granted, we would rhapsodise about their pale, ghostly beauty, the delicate patterning of their upper wings and the subtle yellow-green-black tints of their underwings. We would feel a surge of excitement when we saw them; we would look at them with close attention; we would appreciate them.

Questions and Answers

When people discover that you are interested in butterflies, they will either gulp nervously and move the conversation swiftly along, or they will express an interest (often quite genuine) by asking some pretty basic questions. It is as well to be ready with the answers, so here is a quick briefing on the kind of things you are likely to be asked.

Q: Why are they called 'butterflies'?
A: No one really knows. There are many theories, some more plausible than others. One of the more attractive is that butterflies are named after the butter-yellow (or, in the male, sulphur-yellow) Brimstone, often the first butterfly to be seen in spring. Others have suggested that the name was inspired by butterflies' apparent interest in buttermilk, which does seem to attract some butterflies' attention if it's exposed in the open. There was a folk belief that witches took the form of butterflies and helped themselves to milk and butter. However, even if a butterfly devoted its whole life to consuming dairy products, it would scarcely manage more than a few thimblefuls. Butterflies don't eat at all – they're not physically equipped for it; they only drink. All their eating is done at the larval stage – which makes it even stranger that some sources suggest the possibility (based on an Old Dutch word, 'boterschijte') that butterflies are so named because their excrement resembles butter. Butterflies don't excrete or egest anything, except sometimes a little water if they're overfull, so that theory sounds wildly fanciful. For myself, I favour the confident definition in Johnson's Dictionary: 'A beautiful insect, so named because it first appears in the beginning of the season

for butter.' As the butter season (in the eighteenth century) extended roughly from March to September, this seems to make good sense. But the fact remains that, really, no one knows.

Q: What is the difference between butterflies and moths?
A: Another tricky one, as there are no entirely hard and fast distinctions. However, all butterflies fly only in daytime, whereas most moths fly by night. Most butterflies have relatively slim bodies, while most moths are fat and furry. At rest, most butterflies settle with their wings folded above their bodies, while moths either spread their wings or fold them along their bodies. The one (almost) sure way to tell butterflies and moths apart is that butterflies' antennae always end in a club-like shape, and moths' antennae never do. But that is not terribly useful information for a beginner. Suffice to say that, once your eye is in, you will soon learn how to tell the one from the other – often by the way they fly – ***nearly*** all the time: there are some day-flying grassland moths whose mission in life seems to be to bamboozle the butterfly watcher into thinking they are butterflies.

Q: How do butterflies get through the winter?
A: They hibernate, or become dormant, either as adult butterfly (imago), chrysalis (pupa), caterpillar (larva), or in the egg (ovum). This is why adults emerge at different times across the flying season. The butterflies that hibernate as adults are relatively

large and brightly coloured – Peacock, Tortoiseshell, Comma, Brimstone – and they are the first to appear in spring, or even on mild winter days. The last butterfly to appear, the Brown Hairstreak, overwinters in egg form, as do several others. At least one species, the lovely Speckled Wood, can spend the winter as either caterpillar or chrysalis – a fact that has no doubt contributed to the species' successful spread in recent decades. Even in my lifetime, the Speckled Wood has gone from being a summer butterfly of woodland rides to a ubiquitous all-year-rounder perfectly happy in suburbia.

Q: How does a caterpillar become a butterfly?
A: The process of transformation known as metamorphosis is one of the wonders of nature. The caterpillar, growing as it feeds, is obliged to moult (shed its skin) several times in its life, and after the last moult it reveals a new, very different skin, the hard carapace of a chrysalis. Inside this casing (which is often beautifully marked and shaped), virtually all the body tissue and nervous system of the caterpillar dissolves into a kind of living soup, from which gooey mess the living butterfly is somehow assembled, and in due course emerges from its pupal home. At first its wings are small, soft and baggy, but they harden and grow as they are pumped full of blood (or the butterfly equivalent, known as haemolymph). The crawling caterpillar has become a butterfly, ready to fly – and if it's a female, it had better take to the wing as soon as it can, to avoid lust-crazed males.

Q: How long do butterflies live?
A: It depends on the species. Those that hibernate as adults have the longest lifespan, though much of it will be spent in a dormant state. Generally speaking, the smaller the butterfly, the shorter the life span, a matter of days or weeks rather than months. The overall average life span for British butterflies is probably around two weeks or less. The stubborn folk belief that butterflies live for only a day is false, but some of our smallest species might be on the wing for only a few days.

Q: What do butterflies eat?
A: The short answer is sugar in liquid form, obtained either from nectar or from 'honeydew' exuded by aphids on the leaves of trees. Late-flying butterflies are also partial to rotting fruit, and many species obtain useful minerals from muddy puddles, animal droppings, blood, sweat and other unlikely sources. Butterflies have no mouth parts and obtain all their nourishment through the proboscis that they carry coiled under the body like a watch spring when not in use. Nourishment enters the proboscis through tiny holes and rises by capillary action – no sucking required. The British butterfly with the most depraved appetite is the big and extravagantly beautiful Purple Emperor, whose feeding habits I have written about earlier.

Q: What do butterflies do in rainy weather?
A: Wind and rain are great enemies of butterflies, and few will fly in anything more than a breeze or a spot

of drizzle. In rain, a butterfly, particularly a small one, can quickly become sodden and unable to fly, so it will always seek shelter, often before the rain has started. Hanging under leaves or flower heads with wings folded is usually the best option for smaller species, while others will need the shelter of thick herbage, hollow trees, caves, outbuildings, anywhere reliably protected from the rain. Butterflies know well how to keep dry; in Britain they need to.

The Butterflies, and Where to Find Them

So – butterfly country: what is it, and where can it be found? There are certain habitats that tend to be particularly rich in butterflies, if properly managed: chalk downland kept largely free of scrub, woodland with clearings, coppices and rides, heathland (for certain habitat specialists) and unimproved, flower-rich grassland. There are also, more relevantly to those of us living in towns and suburbs, other, more immediately accessible habitats: gardens, if planted with good, nectar-rich flowering plants and not over-manicured, and other forms of 'edgeland'. Britain, a land where the boundaries between town and country are often blurred, is rich in edgelands, a kind of unofficial countryside that is the sum of huge numbers of patches and strips of neglected, unattractive, easily overlooked land, neither town nor country but a kind of unstructured no man's land between the two. It could be said that the whole of suburbia is an extended edgeland; it is certainly of growing importance to wildlife in various forms, from urban foxes and hedgehogs, gulls and corvids to some of our more adaptable butterflies. Classic examples of edgeland habitats are railway cuttings and other undeveloped land beside railways; motorway verges and unmown roadsides; building and demolition sites and other 'brownfield' land; allotments, if not too tidy; the more neglected areas of parks and gardens; and all the odd patches of land that seem simply to have been forgotten, even by the developers, and reclaimed by nature. This is what the writer and environmentalist Roger Deakin aptly described as 'the undiscovered country of the nearby', and there is a lot of it. With so much of our wildlife having been driven out of the 'real' countryside by intensive farming, these edgelands have become increasingly important as habitat, and butterflies have been among the beneficiaries.

Many edgeland plants, thriving on poor or almost non-existent soil, are nectar-rich and attractive to butterflies. Buddleia is a prime example: this shrub, native to China, has proved itself very much at home in this country, where its seeds, produced in huge numbers, spread initially along railway lines, helped on their way by the draught from passing trains. A plant that thrives in the thinnest and poorest of soils, it is now everywhere, even growing out of walls, roofs and chimneys, and whenever it is in flower it attracts butterflies, especially Whites and those showier species, the Red Admiral, Painted Lady, Tortoiseshell, Peacock and Comma. Because it is so abundant on waste ground and other unpromising patches of land, Buddleia makes such places of interest – or at least worth a look – to the butterfly watcher. We might be surprised by what we find when we look closer: many wild flowers happily colonise such 'waste ground', attracting a range of insect species, including butterflies. Brambles, ivy and traveller's joy, thistles, oxeye daisies, ragwort, knapweed, hemp agrimony, dandelion, red valerian, hawkweeds and Michaelmas daisies are among the plants that spring up in such neglected places, and all are attractive to nectaring butterflies. Wherever there is sunlight and a range of nectar-rich plants, however unprepossessing the site, there you are likely to find butterflies – but only if you look.

When I was working in a London office, my morning commuter train passed a stretch of railway land where two lines meet and a red signal often brought us to a brief halt. It was a scrubby patch, much of it overgrown with brambles, but with some clear grassland, bright with oxeye daisies in summer. I would always look out of the window to see if anything was flying, and was rewarded, over the years, with the sight of more than a dozen butterfly species, including bright little Orange Tips and sulphur-yellow Brimstones in spring; Meadow Browns, Gatekeepers and Blues in summer; and, happiest surprise of all, a small colony of beautiful Marbled Whites right by the railway track. If I had been able to get out and explore, I'm sure I would have found more species, but to see so many from my train seat, simply by looking out of the window, was reward enough, and on a good butterfly morning I always arrived at work with a lighter heart.

These were not the only butterflies I have seen while in transit: Brimstones, Whites, Speckled Woods, Red Admirals and more are often to be seen beside railways, especially on the sunny side of cuttings, and the Brimstone at least is so distinctive that it can be identified even from a speeding intercity train. My best and most unlikely 'spot', though, was from a car window as we waited on a slip road to exit South Mimms motorway services. In an unmistakable flash of yellowish orange, a male Clouded Yellow flew past at speed, on its way to who knows where. It was the only one I saw that year.

The lesson of all this is clear: wherever you are, look about you. 'Butterfly country' is not far to seek: in some form, however small-scale, however attenuated, it is all around us, even in town, even in the city. The heart-lifting surprise of a butterfly encounter might be around any corner. Take in your surroundings, however unpromising they might seem, be aware of sunlight and shade, of where the wind is, seek out the sunny, sheltered spots and the nectar-rich flowers, sharpen your gaze and focus your attention, learn to stand still. You don't know what is there until you look – really look. Pay attention.

What follows is not a list of all the British butterfly species, but rather of those that might be seen, without major excursions, by anyone with some more or less butterfly-rich terrain at not too great a distance – which means most of us not living in the cooler, more mountainous parts of Britain.

Once you start looking, at the right times and in the right places, you should be able to spot twenty or so species, and if you live in one of the more favoured parts of the country, maybe thirty or more.

What follows is also not any kind of field guide, but more a set of thumbnail sketches. I have arranged the butterflies roughly in the order in which they appear through the season, and sketched individual species in a way that I hope will make them easier to find and identify. Distinguishing between species does not present too many difficulties with our British butterflies (unlike their continental cousins), though with some it takes practice, especially if they refuse to settle and allow a close look. I have omitted very localised species, such as the Lulworth and Chequered Skippers and the Swallowtail, and hard-to-find habitat specialists such as the Wood White and the Duke of Burgundy, and have excluded real rarities, such as the Camberwell Beauty and the Large Tortoiseshell.

In general terms, the farther south and west you are in Britain, the more varied and abundant the butterfly life around you is likely to be, especially where there is well-managed woodland and unimproved downland, flower meadows and heathland. There are, however, some northern specialities, such as the Large Heath and the very local Mountain Ringlet, and many species formerly restricted to southern parts have been spreading northward in recent years (and sometimes, as with the Comma and Speckled Wood, for much longer). As I have shown in earlier chapters, even if you are mostly confined to urban and suburban environments, there are still butterflies to be seen, if you keep your eyes open.

I have concentrated on the butterfly itself, rather than its earlier life stages, though sometimes larval food plants are mentioned, if they indicate promising terrain. As I have mentioned before, if you are at all serious about butterfly watching, you will need a good field guide in your pocket or your bag. What follows is no substitute, but will, I hope, serve as a readable introduction to British butterflies, and where and when you might see them.

Butterflies of Spring and Early Summer

Of the early butterflies, many fly on into the later summer or autumn, or reappear later in the season, after a second brood has hatched. The species you are most likely to see in the spring are these, beginning with the emerging hibernators, which sometimes make a fleeting appearance even on a mild winter day:

The Brimstone: The first Brimstones are one of the most cheering sights of the butterfly season, and invariably one of the first. The males, newly awakened from hibernation, are the first to fly, staking out territories in hedgerows and woodland edges and on railway embankments. With their strong sulphur-yellow upperwings and distinctive, leaf-shaped wing profile, they are unmistakable. The females, much paler in colour, can be mistaken in flight for Large Whites, but once settled they are easily identified. Common throughout southern England, the Brimstone has also spread across most of the north in recent times, encouraged by the popularity of buckthorn, its food plant, as a garden and hedgerow shrub. One of the longest-lived of our butterflies, it has only one brood, emerging in high summer, yet can be seen on the wing from early spring to autumn (when it is particularly partial to ivy flowers), with a lull before the summer emergence.

The Small Tortoiseshell: Though much less common than it used to be (thanks to a parasitic fly targeting its caterpillars), this is still one of the UK's most familiar and well-loved butterflies, widespread across the whole country and often turning up in gardens. Its prettily patterned orange upperwings are bordered with blue crescents along scalloped wing margins, while the underwings are dark and bark-like. The caterpillars feed on nettles and the butterflies on a wide range of flowers. During winter they hibernate, often in or near buildings, including garden sheds and church interiors. Awakening in early spring warmth, Small Tortoiseshells are on the wing until late autumn.

The Peacock: The four large, peacock-like eyes on its scalloped, brick-red upperwings make this one of our most spectacular and easily recognised butterflies. The underwings, by contrast, are very dark, nearly black. Like the Small Tortoiseshell, it is common and widespread, feeds on nettles (at the larval stage) and nectar-rich flowers, is often seen in gardens, and hibernates in dry, sheltered places, awakening in early spring and flying into the autumn. There is a lull between the first brood and the second, which emerges in July and sets about feeding up on nectar, ready for hibernation.

The Comma: Its ragged wings, suggesting a victim of bird attack, make the Comma unique among British butterflies. In flight, with its dark-spotted orange upperwings, it might be mistaken for a Fritillary, but once settled with its wings folded, it can be mistaken only for a dead leaf, so perfect is its dark protective coloration, punctuated by the white comma-shaped mark that gives it its name. It has hugely increased its range and numbers since the 1910s, when it seemed to be heading for extinction. Now it is to be found in woodlands and almost any sunny space over most of the country. Another hibernator, it flies from early spring to late autumn, with a mid-season lull before the second brood butterflies emerge.

The Red Admiral: This fascinating butterfly has made more than one appearance in this book already, and, with its distinctive scarlet and black patterning, it will be familiar to anyone who has taken any notice of butterflies, especially as in most years it is abundant and widespread across the whole country. A powerful flier, it migrates from as far south as the Mediterranean coast, arriving from May onwards. Although it doesn't go into hibernation and can survive only the mildest of British winters, there have been years when it has been seen somewhere in the country on every single day. It can be found wherever there are flowers to feed on, and nettles to lay its eggs on, and is a frequent visitor to gardens, especially in autumn when it gorges on late nectar and overripe fruit.

The Painted Lady: The Painted Lady has intricately patterned dull orange wings and cannot easily be mistaken for any other species. It is a migrant which rarely survives the English winter, and its origins are even further south than the Red Admiral's, in North Africa. Numbers fluctuate hugely, but in a 'Painted Lady year' (they happen roughly once a decade) the butterflies spread across the whole country in huge numbers and can be seen on practically any warm, sunny open space. They are especially fond of thistles, both for feeding and egg-laying. The first arrivals usually turn up in the spring, numbers build to a peak in late summer, and individuals can be seen well into the autumn.

The Speckled Wood: This beautiful and common butterfly, with its chocolate-brown upperwings dotted with creamy yellow patches and bright little eyes, has made more than one appearance in this book. Essentially a woodland butterfly, the Speckled Wood is a happy case of a once scarce butterfly that has prospered in recent decades, and can now be found over most of the country and in almost any habitat, including suburban parks and gardens. It can hibernate either as caterpillar or chrysalis, with the result that it can be seen at any time from early spring to late autumn.

The Small White: There are three British species of White, all of which tend to get conflated into the mythical 'Cabbage White'. The first to appear is the Small White. Considerably smaller than the Large White (see below), and less conspicuously marked, the Small White fluctuates in numbers from year to year, but is never less than very common, and in recent years has tended to outnumber the Large. It flies for longer too, emerging as early as February in a mild year, and flying as late as November. In some years, unusually small Large Whites appear, which can lead to confusion, but the Small White usually has a black spot in the centre of each forewing, and the dark tips of the upperwings do not extend down the outer edge of the wing, as they do in both of the other two Whites – the Large and the Green-Veined.

The Holly Blue: This pretty little silvery-blue butterfly has a different way of life from the other Blues, typically flying around shrubs and trees rather than on grassland, and turning up in a wide range of urban and suburban settings, including gardens. The male is paler and smaller than the Common Blue, and larger and bluer than the scarce Small Blue, with similarly speckled pale blue underwings. The female has attractive dark outer edges to its wings, and looks darker in flight. The Holly Blue settles more often on leaves than on flowers, and its foodplants are the ever-abundant ivy and holly. With two and sometimes three (in the south) broods, it can be seen on the wing any time from early spring to late autumn, and happily it has become considerably more common and widespread in recent times.

The Orange-Tip: One of the prettiest and most cheering of the spring butterflies, this species is named for the vivid orange tips of the male's upper forewings (the females have only a grey edging). In both sexes the hindwings are beautifully marked with a green and white mottling that offers perfect protective coloration when the butterfly is perching on a plant with its wings closed. This mottled pattern is faintly visible on the upper side, much stronger on the underside, and it makes the Orange-Tip female quite easily distinguishable from the Small and Green-Veined White. Fairly common through most of Britain, the Orange-Tip can be found in a wide range of habitats, wherever its food plants, notably Cuckoo Flower and Jack by the Hedge, grow. It flies from early April to June.

Harder to find, but worth seeking out, are these habitat specialists:

The Grizzled Skipper: The smallest and most moth-like of all the Skippers – and hence easily overlooked – this species is far from common, but is worth looking out for, more in hope than expectation, between April and June in southern England. It favours rough but quite short-cropped grassland, woodland rides, even the edges of railway tracks, if hot and sunny enough. In flight, the Grizzled Skipper is a greyish blur, easily taken for some nondescript day-flying moth, but when settled, with its wings spread flat – dark brown spangled with white and with delicate chequered fringes – it shows its beauty. It lives in small colonies, and the males are fiercely territorial, engaging in lively high-speed dogfights.

The Dingy Skipper: The name is a little harsh, but this small brown skipper is certainly far from showy, and towards the end of the season it can look decidedly dowdy. It is a little larger than its Grizzled cousin, with less defined markings and no spangles, but it gives a similar impression of a greyish blur in flight, and can easily be taken for a day-flying moth, even when at rest. Like the Grizzled, it settles with wings spread flat on the ground, and favours warm, sunny spots on quite short-cropped grassland. Abandoned quarries and coastal undercliffs are also good sites. Flying from mid-April to the end of June, sometimes with a second brood taking to the wing in July or August, the Dingy Skipper is more common and more widespread than the Grizzled, its range extending further north, but is even easier to overlook, so it is always worth investigating those greyish blurs as they whizz by.

The Green Hairstreak: This little butterfly, which typically flies from late April to June, is rather nondescript in flight, but once it lands and folds its wings, the beautiful emerald green of its underwings identifies it immediately (and, if it settles amid greenery, also renders it hard to spot again). The males are especially lively, chasing females and seeing off intruders before returning to the perches from which they guard their territory. Much less common than it used to be, the Green Hairstreak is still quite widespread and always worth looking for on down and heath, by woodland edges and on embankments. Like all the Hairstreaks, it is easily overlooked.

The Small Blue: 'Small' is the word for this tiny Blue, the smallest of all British butterflies. 'Blue', on the other hand, is less descriptive, as the upperwings of the female are dark brown, and of the male a kind of smoky slate-black with only a dusting of blue. The underwings are silver-grey, lightly sprinkled with black dots, and in flight it has a silvery look. Though widely scattered, it is only at all abundant on some southern chalk and limestone hills and downs, where it is locally common. It lives in small colonies centred on its food plant, Kidney Vetch, and flies in May and June, with a second brood often taking to the air in August.

The Wall Brown (or Wall butterfly): In flight, this brightly coloured butterfly can be mistaken for a Fritillary or even a Comma, but once settled it is unmistakable – and it often settles with wings open to bask. Its orange upperwings are patterned with darker lines and borders, and it has one beady 'eye' at the corner of each forewing, with four smaller ones edging the hindwing. Once common on rough grassland across most of the country, it is now more sparsely distributed and easiest to find on sunny open sites near the coast. Having two broods, it can be seen from May to October.

The Adonis Blue: The male Adonis Blue is spectacularly beautiful, its upperwings of an intense, almost turquoise blue which distinguishes it from even the brightest Common Blue (as do the dark veins running through its white wing margins to the edge of the wing). The female's upperwings are brown, and it can easily be confused with the female Chalkhill Blue. Both species are limited to southern downlands, where the Adonis favours even warmer and sunnier spots than the Chalkhill. Over recent decades, the Adonis Blue has recovered in numbers and spread, but is still scarce and only locally abundant. It is worth seeking out, in likely locations, in late spring and early summer, and again in August and September.

The Pearl-Bordered Fritillary: This pretty little butterfly and the very similar Small Pearl-Bordered Fritillary are now almost entirely confined to western parts, where the Small is the more abundant of the two. The Pearl-Bordered, once a common woodland butterfly, has suffered a steep decline in population as a result of the loss of coppices and open space in woods. The name derives from the pearl-like edging of the mosaic-pattern underwings. Both species fly in late spring and early summer.

These more common butterflies can be seen from spring into high summer or autumn:

The Large White: The largest of our three Whites is very common, and familiar as a voracious garden pest to those who grow brassicas. It is easily taken for granted, but is of great scientific interest for its curious relationship with parasitoid wasps, its use of pungent mustard oils, derived from its food plants, to deter predators (a trait it shares with the Orange-Tip), and its migratory flights. Indeed, the Large White can be a surprisingly strong flier when bent on getting from one place to another, though the rest of the time it is much slower and floppier. It can be seen any time from April to September, with a lull around midsummer.

The Green-Veined White: The third 'Cabbage White' is a delicate and attractive species, which favours damp grassland and woodland rides, and, though not as common as its Small and Large cousins, it can be seen right across the British Isles. Its flight is quite weak and fluttering, and it is typically on the wing from April to September, with a lull around midsummer. It can be told from the similarly sized Small White by the distinctive dark edging of the veins of its underwings – more black than green – which also lend the butterfly a generally darker appearance than the Small.

The Small Copper: This beautiful little butterfly is Britain's only Copper (the spectacular Large Copper having been extinct for a century and a half). With its bright copper-coloured wings, lightly dotted with black, it is conspicuous and unmistakable, very active but happy to settle and bask. Fairly common and widespread across Britain, it is to be seen on any kind of rough grassland where its food plant, sorrel, is growing. Having a second brood, and in warm summers a third and even a fourth (in the south), it is on the wing from April to October, one of the most cheering sights of both spring and autumn.

The Common Blue: Though not as common as it once was, this is indeed our commonest Blue, and the most widely distributed, but it is a butterfly of uncommon beauty. The bright violet-blue, silver-fringed upperwings of the male make it a glorious sight, second only to the Adonis Blue, and the smaller females, with dark upperwings similar to the Brown Argus – but always showing some blue – are differently beautiful. The underwings of both sexes are patterned with white-rimmed black dots and fringed with orange dots. The Common Blue can be found almost anywhere its food plant, Bird's Foot Trefoil, grows, and often turns up in suburban gardens, on waste ground and edgelands. It flies from May to October.

The Brown Argus: Technically a Blue, the Brown Argus has no blue about it, having brown upperwings fringed with orange crescents, and patterned underwings similarly edged with orange spots. Despite its very different coloration, this pretty little butterfly can be confused with the Small Blue in flight, as it gives a similar silvery effect. Once settled, however, it is easily identified. In decline through much of the twentieth century, it turned a corner in the 1990s, and is now widespread across most of southern and eastern England, turning up on rough grassland, downs and woodland rides. It flies from May to September.

The Small Heath: This pretty little butterfly always settles with its wings closed, showing a forewing of orange dotted with a beady eyespot, over a grey-brown hindwing. The upperwings are of tawny gold, but they are rarely on show. Widespread, if not abundant, across almost the whole country, it can be seen on unimproved grassland, banks, verges and woodland rides from May to September. Its larger, darker cousin, the Large Heath, is restricted to northern parts, and is common only in parts of northern Scotland.

The Meadow Brown: One of the most common of all British butterflies, this large Brown can be seen in almost any part of the country, wherever there is more or less wild grassland. Its upperwings are dull brown, the male's marked only by an eyespot on each upperwing and a faint suggestion of orange, the female's much more orange, with more obvious eyespots. It varies quite widely in both size and markings. A weak flier, the Meadow Brown can be seen on the wing at any time from late May to September, even on overcast days.

The Large Skipper: As the name suggests, this Skipper is considerably larger than the Small (see below), but it is large only by Skipper standards. While numbers of both Small and Large fluctuate, the Large is a generally common and widespread butterfly of rough grassland, especially in southern parts. It has a distinctive orange-and-brown mottled pattern on its upperwings, and shows some green on its underwings (as does the Small, but to a less noticeable extent). The Large Skipper's flight season begins and ends earlier than the Small's, typically lasting from mid-May to early August. The males are belligerently territorial, perching on favoured vantage points and patrolling their territory tirelessly. A fast and agile flier, the Large Skipper, when settled, folds its forewings in a V shape and spreads its hindwings flat.

Butterflies of High Summer

The Small Skipper: This pretty little golden-brown butterfly is usually quite common wherever there is rough grass, so it can be seen even in the most unpromising edgeland habitats, and in the less kempt areas of parks and gardens. Quick and extremely mobile in flight, it is not easy to follow, but when settled it folds its forewings into a V shape, while its hindwings are held flat to the ground. Flying from mid-June to late August, it lives in self-contained colonies and seldom wanders far from its home territory. Though far from showy, the Small Skipper is a charming ornament of the summer. The very similar Essex Skipper is also quite common in southern counties, and the two species often fly together. Differentiating between them is an interesting exercise, but perhaps not for the beginner.

The Clouded Yellow: Unlike most Whites, the Clouded Yellow is a strong flier, covering hundreds of miles to reach our shores, and returning south at the end of the season. Numbers fluctuate greatly from year to year, so it is hard to generalise about its abundance, but you are certainly more likely to spot one in the southern counties than further north. When seen it is unmistakable, its dominant colour being a rich, almost orange yellow (in contrast to the Brimstone's more lemon-toned yellow). It also has conspicuous black edgings to its wings. You are more likely to see it in flight than settled, but the delicately marked, greenish-yellow underwings, with an 'eye' unusually placed near the centre of the hindwing, are particularly beautiful.

The Dark Green Fritillary: This beautiful large Fritillary (second in size only to the Silver-Washed) is quite widely distributed on unimproved grassland, especially in the south, and sites all around the western coast. A strong flier, it gets its slightly inappropriate name from its pale greenish underwings, beautifully marked with silver spots of various shapes and sizes. The upperwings are of orange marked with black, with a strongly patterned border. It can be seen on the wing from June to August, and is a glorious sight, well worth seeking out.

The Marbled White: Although it is technically a Brown, its name is perfectly apt for a butterfly whose upperwings are chequered black and white and resemble the patterning on marble – a resemblance even more marked in the paler, more delicately patterned underwings. With their dancing flight and readiness to settle on flowerheads (especially purple flowers), these unmistakable butterflies are one of the loveliest sights of summer, and happily, as their range spreads eastward and northward from southern England, they are becoming quite common in places where they were not seen before. A butterfly of grassland and woodland rides, the Marbled White flies from June to August.

The White Admiral: More than once in the course of this book, I have rhapsodised over the beauty of this butterfly, the lovely black-brown and white of its upperwings, the exquisitely patterned bronze and white of its underwings, its supremely elegant flight, in and out of sun and shade on woodland rides. One of the few species to have benefited from changes in woodland management (it likes a mix of shade and light), it has extended its range, but is never more than locally common in large woods in southern counties, where it flies from mid-June to August.

The Silver-Washed Fritillary: The largest of our Fritillaries, and the most spectacular, this is a woodland butterfly that, unlike its Pearl-Bordered cousin and many other species, has not been shaded out by overgrowth. A strong flier, it is quite widespread in southern and western woods, where it makes an impressive sight flying in sunny rides and glades, but it also turns up in unexpected places, even gardens. The male is a brighter orange than the female, and slightly smaller. The upperwings of both sexes are marked with black dots and lines, and the hind underwings are green (more green, in fact, than the Dark Green Fritillary), streaked with delicate silvery lines. It is on the wing from late June to September.

The Purple Emperor: This butterfly, in size second only to the Swallowtail, has an alure all its own, being elusive and unpredictable, 'imperious' in its behaviour, and spectacularly large and beautiful. It spends most of its life high up in oak trees, but when seen, even in flight and at a distance, it is hard to mistake, thanks to its size and vigour, and, when seen close up, the purple sheen on the male's upperwings and the oddly smeary appearance of its banded and eyed underwings make it impossible to mistake for anything else. Not as rare as is widely thought, but still scarce, it can be found (with luck) in woodland on clay soils in southern counties, and has been gradually extending its range northward. It flies from late June to August.

The Gatekeeper: The lively little butterfly gives a bright orange impression in flight, the result of the large orange patches that cover most of its brown upperwings. These are dotted with eyespots that look unusually beady, having a double 'pupil', and the underwings are prettily banded and eyed. With its sporadic, dancing flight and love of settling to feed on bramble and other flowers, the Gatekeeper is a happy and familiar sight of summer. Common on grassland, woodland edges and hedgerows as far north as the Midlands, it typically flies from late June to early September.

The Ringlet: With its upperwings of rich, dark brown and the string of black eyespots, haloed with yellow, that decorate its underwings, this is a quietly beautiful butterfly. It is a joy to watch in flight too, as it dances over a grassy flower meadow or settles to join the Meadow Browns and Gatekeepers feasting on bramble flowers. Quite common on grassland and woodland edges in eastern and southern parts, it favours long grasses and, unlike most of our butterflies, benefits from a wet summer making its food plants more luxuriant. It flies typically from late June to late August.

The Purple Hairstreak: This is the most common and widespread of our Hairstreaks. It is a butterfly of oak woodlands, but even a single oak tree on a suburban street can support a small colony, so it's always worth looking out for this one. In flight it can be, like the Green Hairstreak, quite nondescript, but when the light catches it, a glint of purple and silver makes it unmistakable. It spends most of its life high up in oak trees, flying about or resting, but when it has settled in view, it can be identified by the delicate markings on the silver-grey underwings – a wavy white streak and an orange eye beside the tail – and by the subtle sheen of purple on its upperwings. The Purple Hairstreak flies in July and August.

The White-Letter Hairstreak: Every bit as elusive as the Brown, this is the darkest of the Hairstreaks and perhaps the most easily overlooked. It spends most of its life high up in elm trees (of any species, but most commonly Wych Elm) and rarely descends to lower levels. When seen close up, with its wings folded – as they invariably are – it can be identified by the distinctive jagged streak on the underwing, ending in a sideways W (hence 'White-Letter') beside the tail. Though always local, it is widespread across most of England and Wales, and might well be much more numerous than is thought. Like the other tree-dwelling Hairstreaks, it is often best looked for with binoculars. It flies, like the Purple Hairstreak, in July and August.

Butterflies of Late Summer and Autumn

Most of the butterflies to be seen late in the season have been on the wing earlier in the year, but there are also a few late arrivals at the butterfly ball:

The Chalkhill Blue: As its name suggests, this large and handsome Blue is a butterfly of chalk and limestone downland, and it is now found only in southern England. The male's upperwings are of pale blue, varying in shade from silvery to milky, with dark margins, while the female's upperwings are a rich brown. The underwings are grey or brown, thickly dotted. The Chalkhill Blue can be locally very abundant, with thousands flying in a good year, and many surviving colonies have benefited from being on protected sites. A butterfly of high summer, it is on the wing from July to September.

The Grayling: The largest of the Browns, this butterfly has pale brown upperwings banded with still paler brown and dotted with two conspicuous eyes on each forewing. However, it rarely settles with its wings open, folding them to show an under hindwing that looks exactly like a piece of bark and makes it all but invisible on the kind of broken ground it favours. Once much more widespread, it is now confined to southern heathland and arid coastal sites where vegetation is sparse. It flies from July to September.

The Silver-spotted Skipper: If you are lucky enough to live near good southern downland, with warm, south-facing slopes, you might also be lucky enough to find this lovely skipper, which is among the latest of our butterflies to emerge, usually appearing in late July or August. In general appearance, it is similar to the Large Skipper, but seen close up, especially with its silver-dotted green underwings on show, it has a beauty all its own. The most heat-loving of all our butterflies, it seldom flies in anything less than full sun or in temperatures below twenty degrees Celsius. As adults only live six days or so, this makes it peculiarly vulnerable to bad summer weather, and it is perhaps no wonder it is something of a rarity, especially as it is reluctant to spread far from its established sites. However, the outlook for the Silver-Spotted Skipper has improved considerably over recent decades, because of warming and the return of rabbits, so vital to keeping downland close-cropped.

The Brown Hairstreak: From the earliest flying Hairstreak to the latest – the Brown Hairstreak is one of the last butterflies of the summer, flying from late July into September. It is also the largest of our Hairstreaks, and in flight it can be confused with the Gatekeeper, but, once settled, its golden underwings, with delicate hairlines and 'tail', are unmistakable. With colonies scattered across the southern counties, it is not a rare butterfly, but extremely elusive, spending much of its life in the treetops (often in an ash tree) and descending to earth only to lay its eggs, and sometimes to bask. Its food plant is blackthorn, and it has, like many other species, suffered from the tidying-up and excessive cutting back of hedgerows. However, it does sometimes turn up in unexpected places, including suburban gardens.

The last butterflies you see in a typical season are likely to be the same as the first – the hibernators, now feeding up on late-flowering nectar-rich plants such as ivy, or on fallen fruit, in preparation for their long winter sleep: Brimstones, Tortoiseshells, Peacocks, Commas. And, taking its chances on surviving the winter to fly again – the Red Admiral.

Coda: Camilla

Juniper Hall, a Victorianised eighteenth-century country house which now serves as a field studies centre, stands on the east slope of the Mole gap in the North Downs, close to Box Hill, amid fine downland habitats – unimproved chalk grassland galore, coppiced woods, heathland, streams and springs. This is some of the finest butterfly country in England, home to such rarities as the Adonis Blue and Silver-Spotted Skipper, and the regional branch of Butterfly Conservation often holds events at Juniper Hall.

The interior is much altered, but one room retains its eighteenth-century elegance – the Templeton Room. It was here that the novelist and diarist Fanny Burney, on a visit to nearby Norbury Park, met her future husband, one of a group of French émigrés who had escaped the Revolution and were living temporarily at Juniper Hall. General Alexandre d'Arblay, a career soldier who had been adjutant-general to Lafayette, greatly impressed Miss Burney, and the two soon became very close. D'Arblay gave her French lessons and introduced her to other Juniper Hall émigrés, including the somewhat scandalous Madame de Staël. He courted Fanny at her parental home in Chelsea, and in July 1793 the couple were married in the parish church of St Michael in Mickleham, the village close to Juniper Hall. They walked to the church from her sister's cottage nearby.

It seems to have been an extremely happy marriage, and a son was born the following year. By this time, Fanny had begun work on a new novel, to be called, after its heroine, Clarinda, or perhaps Ariella – yes, she wrote to her brother, 'the name of my heroine is ARIELLA'. But no, her father (the music historian Charles Burney) objected to that name, and so, shortly before it was published in 1796, her new novel became Camilla. It was the right choice, and Camilla was to prove her most commercially successful publication. On the proceeds, she and d'Arblay built a cottage in the hamlet of Westhumble, just down the road from Juniper Hall and Mickleham – Camilla Cottage. Sadly, the cottage has not survived: it burnt down in 1919 and was replaced by a much larger house, Camilla Lacey, which has its own wide entrance arch in faux Perpendicular style.

The ghost of Camilla Cottage is one point of a Box Hill literary triangle. Down the road from Westhumble, at the foot of the dip slope of Box Hill, is the Burford Bridge hotel, where Keats stayed in 1817, while he was writing Endymion ('I like this place very much. There is Hill & Dale and a little River – I went up Box hill this Evening after the Moon – you a' seen the Moon – came down – and wrote some lines…'). He was not the only literary guest to have stayed there: Robert Louis Stevenson, Jane Austen, Wordsworth and Sheridan all visited. Only a short distance away, across the slope of Box Hill, is a remarkable literary survival – George Meredith's enviably spacious 'writing chalet', in the grounds of his home, Flint Cottage (now a private house, but the chalet is conspicuously visible). Meredith, novelist and poet, was in his day a giant of English letters, but he is sadly little read today. He loved the Surrey downs and often rhapsodised about his surroundings: 'I am every morning at the top of Box Hill, – as its flower, its bird, its prophet. I drop down the moon on one side, I draw up the sun on t'other. I breathe fine air. I shout ha ha to the gates of the world…' The writing chalet is at the top of a steep path from the house, and towards the end of his life, when he couldn't manage the climb, Meredith would be pulled up in a bath chair by a donkey called Picnic. From Flint Cottage it is but a short step to Juniper Hall…

But back to Camilla. Like any bookish young lady of her time, Fanny Burney would have known her Virgil, and was no doubt familiar with the huntress and warrior Camilla in the *Aeneid* – the same Camilla that Linnaeus had in mind when, in 1764, he named the White Admiral *Ladoga camilla* (it is now *Limenitis camilla* – you can never rely on scientific names to stay the same for long). Aurelians, who loved the romance of Latin butterfly names, habitually referred to the White Admiral simply as *Camilla*. Venturing into the realms of speculation, I wonder if Fanny might have heard the name used in that context. There would surely have been a scattering of aurelians haunting the woods and slopes around Box Hill, and *Camilla* would have been among the objects of their desire. It would almost certainly have been more abundant then, but the White Admiral still flies in the woods near Westhumble, and I have often seen it there. In fact it was there that I saw my first White Admiral since those well-remembered days of my boyhood. It felt like a reunion.

The flight of the White Admiral is one of the most enchanting sights to be seen (if you are lucky) in English woodland. As Jeremy Thomas puts it (in *Butterflies of Britain and Ireland*), 'No account can do justice to the White Admiral's dainty movements, or convey the character of a creature so ideally suited to gliding in and out of dappled shade among the branches of mature woodlands.' The Victorian entomologist C.G. Barrett (who once spotted two Clouded Yellows from a train, leapt off at the next station, ran back down the line and caught them both) described the White Admiral's 'special grace', which 'seems to arise from the habit of the insect of sweeping down over the trees to near the ground, then rising a little, gliding into every opening, taking the curves of the branches and high bushes with the perfection of ease, sweeping rapidly away, or soaring over the trees, to return in a few minutes to the same spot.'

Camilla casts a rare spell. In 1803, the lepidopterist A.H. Haworth wrote of 'an old Aurelian of London, so highly delighted at the inimitable flight of *Camilla* that, long after he was unable to pursue her, he used to go to the woods and sit down on a stile, for the sole purpose of feasting his eyes on her fascinating evolutions.' I fancy I see my future self in that old aurelian, sitting by some likely woodland ride and watching the White Admirals in flight, my butterfly life come full circle.

Resource List for Further Reading

Books

Field Guides

Jeremy Thomas and Richard Lewington. *The Butterflies of Britain and Ireland*. Bloomsbury. 1991, last revised 2016.

The definitive guide, and a beautiful book in itself, thanks to Lewington's superb watercolour illustrations. A true modern classic that is unlikely to be surpassed, it is a must for anyone seriously interested in British butterflies.

J.A. Thomas. *Philip's Guide to Butterflies of Britain and Ireland*. Philip's. 2007 (previously published as the *RSNC Guide to the Butterflies of the British Isles* and *The Hamlyn Guide to Butterflies of the British Isles*).

Informative, clearly written and well designed, this is an excellent pocket guide, illustrated with a mixture of photographs and watercolours, and with distribution maps and life cycle charts for each species.

Richard Lewington. *Pocket Guide to the Butterflies of Great Britain and Ireland*. Bloomsbury. Second edition. 2015.

A true pocket guide, even more compact than the Philip's Guide, this is a beautiful little volume, illustrated throughout with Lewington's superb watercolours, and including all the basic information you are likely to need.

David Newland, Robert Still, Andy Swash and David Tomlinson. *Britain's Butterflies: A Field Guide to the Butterflies of Great Britain and Ireland*. Princeton University Press. Fourth edition. 2020.

The ultimate pictorial guide, this fairly compact volume is lavishly illustrated with photographs of every British species in various poses, along with basic information, distribution maps and life cycle charts. Particularly useful for identification.

U.S. based guides

Jim P. Brock and Kenn Kaufman. *Kaufman Field Guide to Butterflies of North America*. Mariner Books. 2006.

Probably the best compact identification guide, with over two thousand images, information-rich text, a pictorial table of contents, and range maps for each species.

Jeffrey Glassberg. *A Swift Guide to Butterflies of North America*. Princeton University Press. Second edition. 2017.

The most comprehensive photographic field guide, with over three thousand photographs, informative text, a visual index and range maps.

Richard K. Walton. *National Audubon Society Pocket Field Guide: Familiar Butterflies of North America*. Knopf. 1990.

A very compact, pocket-sized guide, covering only eighty familiar species, but providing a useful introduction to the subject. The Audubon Society also publishes the much larger and more comprehensive Field Guide to North American Butterflies, by Robert Michael Pyle (2005).

Historical

Michael A. Salmon, with additional material by Peter Marren and Basil Harley. *The Aurelian Legacy: British Butterflies and Their Collectors*. University of California Press. 2000.

This large, handsome and richly illustrated volume represents an extraordinary feat of scholarship and research. As well as a full-scale biographical dictionary of collectors, it includes a fascinating history of the British passion for butterflies, and essays on species of historical interest. The most complete and authoritative history.

Michael A. Salmon and Peter J. Edwards, with illustrations by Tim Bernhard. *The Aurelian's Fireside Companion: An Entomological Anthology*. Paphia Publishing. 2005.

A companion volume to The Aurelian Legacy, and every bit as large and handsome, this treasury of butterfly lore celebrates the passionate enthusiasm of the 'brethren of the net', focusing on the golden age of collecting, and drawing on magazines and newspapers, letters, journals and memoirs to build a rich picture of a lost world.

David Elliston Allen. *The Naturalist in Britain: A Social History*. Allen Lane. 1976. revised edition published by Princeton University Press, 1994.

This highly readable classic account of the rise of interest in natural history in Britain naturally includes a good deal about butterfly collecting, usefully setting it in a wider social context.

Moses Harris. *The Aurelian*. Country Life Books. 1986.

A reprint of the unaffordable 1766 original, with an introduction by Robert Mays and marginal notes illuminating the text. Not only a beautiful book, but also a window into the eighteenth-century world of the Aurelians.

Resource List for Further Reading

General

Peter Marren. ***Rainbow Dust: Three Centuries of Delight in British Butterflies.*** Square Peg. 2015.

Wildlife writer and 'repentant collector' Peter Marren's delightful book, at once scholarly and accessible, is the closest book to the one you are currently reading in subject matter, though it does not range quite so widely across literature and the arts, nor into the philosophical, therapeutic and 'mindful' aspects of butterfly watching.

Patrick Barkham. ***The Butterfly Isles: A Summer in Search of Our Emperors and Admirals.*** Granta. 2010.

This engaging and very readable book chronicles Barkham's quest to see and photograph every British species of butterfly in one summer. It's an eventful journey that takes him all over the British Isles, meeting fellow enthusiasts, and musing about the particular magic of butterflies and their place in his life.

Wendy Williams. ***The Language of Butterflies: How Thieves, Hoarders, Scientists, and Other Obsessives Unlocked the Secrets of the World's Favourite Insect.*** Simon & Schuster. 2020.

This American book, written in a breezy journalistic style, investigates butterflies and the people who work with them across the globe. Divided neatly into Past, Present and Future, it is particularly good on recent scientific research.

Paul Whalley. ***Butterfly Watching.*** Severn House Naturalists' Library. 1980.

The first book written entirely from the point of view of ***watching*** butterflies, this is a useful practical guide to finding them, recording observations, rearing butterflies at home, attracting them to the garden, and (in a section written by wildlife photographer Heather Angel) photographing them.

David G. Measures. ***Bright Wings of Summer.*** Cassell. 1976.

This wonderfully readable introduction to the world of butterflies is made special by the author's deep first-hand engagement with the subject, and by his brilliant, impressionistic watercolour sketches of butterflies in flight and at rest. A handsome and illuminating book – described in a foreword by David Bellamy as 'a perfect amalgam of art and science' – it surely deserves to be reprinted.

David Measures. ***Butterfly Season: 1984.*** Arlequin Press, 1996.

This beautiful volume draws together Measures's butterfly

drawings, watercolours and field notes from the wonderful summer of 1984. Opposite each of the sixty colour plates is a transcript of his hastily scribbled field notes made on the spot. No other book brings the reader so vividly close to the experience of being in the field among butterflies.

Matthew Oates. *In Pursuit of Butterflies: A Fifty-Year Affair.* Bloomsbury, 2021.

In a memoir suffused with enthusiasm, lightly-worn learning and sheer joy, conservationist and butterfly man Matthew Oates (of the *Purple Empire* blog) looks back on a lifetime spent in thrall to the beauty and magic of butterflies.

Online

Butterfly Conservation (www.butterfly-conservation.org)

A very useful and wide-ranging website, with features including the latest butterfly news, an excellent identification guide, information on recording and monitoring butterflies, on gardening for butterflies, on 'Why Butterflies Matter', and more. *The State of the UK's Butterflies 2022*, the most recent comprehensive survey of the subject, is available to download on the website.

The North American equivalent of Butterfly Conservation: the North American Butterfly Association (naba.org)

Promotes awareness of butterfly conservation, and encourages monitoring and recording and butterfly gardening. Like Butterfly Conservation, it organises an annual butterfly count each year and publishes the results.

UK Butterfly Monitoring Scheme (ukbms.org)

A resource for those wishing to get seriously involved in the important business of monitoring butterfly populations. The website tells you all you need to know, as well as publishing latest results.

Garden Butterfly Survey (gardenbutterflysurvey.org)

This site encourages butterfly lovers to register their gardens with the scheme, report sightings, see the latest findings, and get useful advice on butterfly gardening.

Works Cited

Alex Fox. 'Study Reveals the Secrets of Butterfly Flight'. *Smithsonian Magazine.* January 2021.

Alfred Russel Wallace. *The Malay Archipelago.* 1869. Penguin Classics. 2014.

Benjamin Wilkes. *English Moths and Butterflies.* 1749. Salzwasser-Verlag Gmbh. 2009.

'Butterfly Senses'. *Monarch Joint Venture.* Accessed online: monarchjointventure.org/monarch-biology/butterfly-senses

Bruce Chatwin. *Songlines.* Franklin Press. 1987.

Charles Darwin. *On The Origin of Species.* 1859. Oxford University Press. 2008.

Claire Thompson. *The Art of Mindful Birdwatching.* Leaping Hare Press. 2017.

'Darwin and Design' *Darwin Correspondence Project.* 2022. Accessed online: darwinproject.ac.uk/commentary/religion/darwin-and-design

David G. Measures. *Bright Wings of Summer.* Cassell. 1976.

David Measures. *Butterfly Season: 1984.* Arlequin Press. 1996.

E.B. Ford. *Butterflies.* Collins. 1945.

Edward Thomas 'The Brook'. 1918. Accessed online: www.poetryfoundation.org/poems/53749/the-brook-56d2335518e67

Edward Topsell. *History of Four-footed Beasts and Serpents.* E. Cotes. 1658. Accessed online: archive.org/details/historyoffourfoo00tops/page/n9/mode/2up.

Eleazar Albin. *A Natural History of English Insects.* William Innys. 1749. Accessed online: https://www.rct.uk/collection/1057018/a-natural-history-of-english-insects

Elizabeth Gaskell. *Mary Barton.* 1848. Penguin Classics. 2003.

Emily Conover. 'Butterfly-inspired nanostructures can sort light'. *Science News.* June 2016. Accessed online: www.sciencenews.org/article/butterfly-inspired-nanostructures-can-sort-light.

Emily Dickinson. *The Complete Poems.* Faber & Faber. 2016.

Erasmus Darwin. *The Botanic Garden.* 1791. Accessed online: www.gutenberg.org/cache/epub/10671/pg10671-images.html

The Butterfly

Erich Maria Remarque. *All Quiet on the Western Front*. 1928. Vintage Classics. 2005.

Friedrich Nietzsche. 'The Wanderer and His Shadow'. 1880. *Human, All-Too Human Part II*. Accessed online: www.gutenberg.org/files/37841/37841-pdf.pdf

Geoffrey Hill. *The Orchards of Syon*. Penguin Books. 2002.

H.G. Knaggs. *The Lepidopterist's Guide*. J. Van Voorst. 1869. Accessed online: *https://archive.org/details/lepidopteristsgu00knag_0*

Henry David Thoreau. 'Walking'. *The Atlantic*. 1862. Accessed online: www.theatlantic.com/magazine/archive/1862/06/walking/304674/

Iris Murdoch. *The Sovereignty of Good*. 1970. Routledge Classics. 2nd edition. 2001.

Isaac Newton, *Principia*. 1687. Prometheus. 1995.

James Petiver. 'Papilionum Britanniae Icones…'. 1767. Accessed online: www.biodiversitylibrary.org/part/297418.

J.A. Thomas. *Philip's Guide to Butterflies of Britain and Ireland*. Philip's. 2007 (previously published as the *RSNC Guide to the Butterflies of the British Isles* and *The Hamlyn Guide to Butterflies of the British Isles*).

Jennifer Frazer. 'Butterflies in the Time of Dinosaurs, with Nary a Flower in Sight'. *Scientific American*. July 2016. Accessed online: *www.scientificamerican.com/blog/artful-amoeba/butterflies-in-the-time-of-dinosaurs-with-nary-a-flower-in-sight/*.

Jeremy Thomas and Richard Lewington. *The Butterflies of Britain and Ireland*. Bloomsbury. 1991. Last revised 2016.

John Clare. 'To the Butterfly'. 1820. Accessed online: allpoetry.com/To-A-Butterfly

John Ray, et al. *Historia Insectorum*. A. & J. Churchill. 1710. Accessed online: doi.org/10.5962/bhl.title.10430.

John Ruskin. 'Essay on the Relative Dignity of the Studies of Painting and Music'. *The Works of John Ruskin*. Ed. Edward Tyas Cook and Alexander Wedderburn. Cambridge, Cambridge University Press, 2010. Accessed online: tinyurl.com/555yp8vp

Kay Ryan. 'A Consideration of Poetry'. *Synthesizing Gravity:*

Works Cited

Selected Prose. Grove Press. 2020.

Kingsley Amis. 'To H.' *Memoirs*. 1991. Vintage Classics. 2004.

L.C. Johansson, P. Henningsson. 'Butterflies fly using efficient propulsive clap mechanism owing to flexible wings'. *Journal of the Royal Society*. Volume 18, Issue 174. January 2021.

Laetitia Jermyn. *The Butterfly Collector's Vade Mecum, or, A Synoptical Table of English Butterflies*. J. Raw, Hurst, Rees & Co. and G. & W.B. Whittaker, London, 1824. Accessed online: doi.org/10.5962/bhl.title.51417.

Leslie Mertz, Ph.D. 'Butterfly Pupae Make Sounds in Never-Before-Known Ways'. *Entomology Today*. September 2018.

Ludwig Wittgenstein. *Philosophical Investigations*. 1953. Wiley. 2009. Blackwell. 2009.

Maria Sibylla Merian. *Metamorphosis*. 1705. Accessed online: www.rct.uk/collection/1085787/metamorphosis-insectorum-surinamensium.

Matthew Oates. *The Purple Empire*. apaturairis.blogspot.com

Michael A. Salmon, Peter J. Edwards, Tim Bernhard. *The Aurelian's Fireside Companion: An Entomological Anthology*. Paphia Publishing. 2005.

Michael A. Salmon, Peter Marren, Basil Harley. *The Aurelian Legacy: British Butterflies and Their Collectors*. University of California Press. 2000.

Moses Harris. *The Aurelian*. 1766. Country Life Books. 1986.

Patrick Barkham. *The Butterfly Isles: A Summer in Search of Our Emperors and Admirals*. Granta. 2010.

Richard Lewington. *Pocket Guide to the Butterflies of Great Britain and Ireland*. Bloomsbury. Second edition. 2015.

Richard Mabey. *Nature Cure*. Vintage. 2008.

Richard Wiseman. *The Luck Factor*. Arrow. 2004

Richard South. *The Butterflies of the British Isles*. F. Warne. 1906. Accessed online: archive.org/details/butterfliesofbri00sout

Robert Hooke. *Micrographia*. 1665. Accessed online: royalsociety.org/ttp/ttp.html?id=a9c4863d-db77-42d1-b294-fe66c85958b3&type=book.

The Butterfly

Robert Louis Stevenson. *Essays of Travel.* Chatto & Windus. 1905. Accessed online: https://www.gutenberg.org/files/627/627-h/627-h.htm

Sana Suri. 'Despite metamorphosis, moths hold on to memories from their days as a caterpillar.' *The Conversation.* August 2014. Accessed online: theconversation.com/despite-metamorphosis-moths-hold-on-to-memories-from-their-days-as-a-caterpillar-29859.

Siegfried Sassoon. *The Old Century and Seven More Years.* Faber and Faber. 1938. Accessed online: archive.org/details/oldcenturysevenm0000sass/page/n5/mode/2up

Simone Weil. *Gravity and Grace.* 1947. Taylor & Francis Ltd. 2002.

Stephen H. Blackwell, Kurt Johnson. *Fine Lines: Nabokov's Scientific Art.* Yale University Press. 2016.

Thomas Browne. *Pseudodoxia Epidemica: Or, Enquiries Into Commonly Presumed Truths.* 1672. Benediction Books. 2009.

Thomas Moffett, et al. ***Insectorum sive minimorum animalium theatrum.*** T. Cotes, 1634. Accessed online: www.biodiversitylibrary.org/page/39877088.

Vladimir Nabokov. *Ada or Ador.* 1969. Penguin Classics. 2000.

Vladimir Nabokov. *Nabokov's Dozen.* 1916. Penguin Classics. 2023.

Vladimir Nabokov. *Speak, Memory.* 1951 Penguin Classics. 2016.

Walt Whitman. *Specimen Days.* 1882. Oxford University Press. 2023.

Walt Whitman. *The Complete Poems.* Penguin Classics. 2004.

William Kirby, et al. *An Introduction to Entomology, or, Elements of the Natural History of Insects : With Plates.* Longman, Hurst, Rees, Orme, and Brown. 1816. Acccessed online: https://doi.org/10.5962/bhl.title.105744.

William Roscoe Thayer. 'Personal Recollections of Walt Whitman'. *Scribner's Magazine*, vol.65, No 6. 1919. Accessed online: https://modjourn.org/issue/bdr479737/

William Wordsworth. 'To a Butterfly'. 1807. Accessed online: allpoetry.com/To-A-Butterfly